D0498581

POLYMER ANALYSIS

Barbara H. Stuart

University of Technology, Sydney, Australia

Baffins Lane, Chichester,
West Sussex, PO19 1UD, England
National 01243 779777
International (+44) 1243 779777

e-mail (for orders and customer service enquiries): cs-books@wiley.co.uk
Visit our Home Page on http://www.wiley.co.uk or http://www.wiley.com

Other Wiley Editorial Offices

John Wiley & Sons, Inc., 605 Third Avenue,
New York, NY 10158-0012, USA

WILEY-VCH Verlag GmbH,
Pappelallee 3, D-69469 Weinheim, Germany

John Wiley & Sons Australia Ltd,
33 Park Road, Milton, Queensland 4064, Australia

John Wiley & Sons (Canada) Ltd, 22 Worcester Road,
Rexdale, Ontario M9W 1L1, Canada

John Wiley & Sons (Asia) Pte Ltd, 2 Clementi Loop #02-01,
Jin Xing Distripark, Singapore 129809

British Library Cataloguing in Publication Data

A catalogue record for this book is available from the British Library

ISBN 0471899267 (Hb)
ISBN 047181363X (Pb)

Typeset in 10/12pt Times by Laserwords, Private Limited, Chennai, India.
Printed and bound in Great Britain by Biddles Ltd, Guildford and King's Lynn.
This book is printed on acid-free paper responsibly manufactured from sustainable forestry in which at least two trees are planted for each one used for paper production.

Contents

Series Preface

There has been a rapid expansion in the provision of further education in recent years, which has brought with it the need to provide more flexible methods of teaching in order to satisfy the requirements of an increasingly more diverse type of student. In this respect, the *open learning* approach has proved to be a valuable and effective teaching method, in particular for those students who for a variety of reasons cannot pursue full-time traditional courses. As a result, John Wiley & Sons, Ltd first published the Analytical Chemistry by Open Learning (ACOL) series of textbooks in the late 1980s. This series, which covers all of the major analytical techniques, rapidly established itself as a valuable teaching resource, providing a convenient and flexible means of studying for those people who, on account of their individual circumstances, were not able to take advantage of more conventional methods of education in this particular subject area.

Following upon the success of the ACOL series, which by its very name is predominately concerned with Analytical *Chemistry*, the *Analytical Techniques in the Sciences* (AnTS) series of open learning texts has now been introduced with the aim of providing a broader coverage of the many areas of science in which analytical techniques and methods are now increasingly applied. With this in mind, the AnTS series of texts seeks to provide a range of books which will cover not only the actual techniques themselves, but *also* those scientific disciplines which have a necessary requirement for analytical characterization methods.

Analytical instrumentation continues to increase in sophistication, and as a consequence, the range of materials that can now be almost routinely analysed has increased accordingly. Books in this series which are concerned with the *techniques* themselves will reflect such advances in analytical instrumentation, while at the same time providing full and detailed discussions of the fundamental concepts and theories of the particular analytical method being considered. Such books will cover a variety of techniques, including general instrumental analysis, spectroscopy, chromatography, electrophoresis, tandem techniques, electroanalytical methods, X-ray analysis and other significant topics. In addition, books in the series will include the *application* of analytical techniques in areas such as environmental science, the life sciences, clinical analysis, food science, forensic analysis, pharmaceutical science, conservation and archaeology, polymer science and general solid-state materials science.

Written by experts in their own particular fields, the books are presented in an easy-to-read, user-friendly style, with each chapter including both learning objectives and summaries of the subject matter being covered. The progress of the reader can be assessed by the use of frequent self-assessment questions (SAQs) and discussion questions (DQs), along with their corresponding reinforcing or remedial responses, which appear regularly throughout the texts. The books are thus eminently suitable both for self-study applications and for forming the basis of industrial company in-house training schemes. Each text also contains a large amount of supplementary material, including bibliographies, lists of acronyms and abbreviations, and tables of SI Units and important physical constants, plus where appropriate, glossaries and references to original literature sources.

It is therefore hoped that this present series of text books will prove to be a useful and valuable source of teaching material, both for individual students and for teachers of science courses.

Dave Ando
Dartford, UK

Preface

Polymers are of major economic and social importance and thus it is necessary to understand the appropriate methods for characterizing such materials. Although there are a number of polymer science texts currently on the market, there are few which are directed at the true beginner to the field. Therefore, one of the aims of this present book is to explain the fundamentals of the subject in a straightforward, clear and concise manner. There are also several books available which cover the specific techniques used to analyse and characterize polymers. This text further aims to introduce the most commonly used techniques for polymer analysis in one book – again, at a level suitable for the beginner.

This text is not intended to be comprehensive – polymer science is a very extensive field! However, it is hoped that the information provided here can be used as a starting point for more detailed investigations. The book is laid out with chapters covering the main aspects of polymer science and technology, namely identification, polymerization, molecular weight, structure, surface properties, degradation, and mechanical properties. The background to each analytical technique is introduced and explained, and how these techniques may be applied to the study of polymers is then covered in the various chapters. Suitable questions and problems (in the form of self-assessment and discussion questions (SAQs and DQs)) are included in each chapter to assist the reader in understanding the specific techniques/analytical methods being discussed.

I should like to thank Kin Hong Friolo and Kristen Nissen for providing data, and, in particular, Paul Thomas for his contributions and his support during the period that this book was being prepared.

Finally, I very much hope that those learning about and researching polymers will find this text both a useful and valuable introduction to the area of 'Polymer Analysis'.

Barbara Stuart
University of Technology, Sydney, Australia

Acronyms, Abbreviations and Symbols

ABS	acrylonitrile–butadiene–styrene
AFM	atomic force microscopy
AIBN	azobisisobutyronitrile
ATR	attenuated total reflectance
DGEBA	diglycidyl ether of bisphenol A
DMA	dynamic mechanical analysis
DMTA	dynamic mechanical thermal analysis
DOP	dioctyl phthalate
DP	degree of polymerization
DRIFT	diffuse reflectance
DSC	differential scanning calorimetry
DTA	differential thermal analysis
EPR	electron paramagnetic resonance
ESCA	electron spectroscopy for chemical analysis
ESEM	environmental scanning electron microscopy
ESR	electron spin resonance
FIB	fast-ion bombardment
FID	free-induction decay
FT	Fourier transform
FTIR	Fourier-transform infrared (spectroscopy)
GC	gas chromatography
GPC	gel permeation chromatography
HDPE	high-density polyethylene
HMA	hexyl methacrylate
HPLC	high performance liquid chromatography
HPP	high-performance polymer

ICP	intrinsically conducting polymer
IGC	inverse gas chromatography
IPN	interpenetrating polymer network
I	initiator concentration
K–K	Kramers–Kronig
KRS-5	thallium iodide
LCD	liquid crystal display
LCP	liquid crystalline polymer
LCST	lower critical solution temperature
LDPE	low-density polyethylene
LLDPE	linear low-density polyethylene
LOI	limiting oxygen index
MALDI	matrix-assisted laser desorption ionization
M–F	melamine–formaldehyde
MMA	methyl methacrylate
MS	mass spectrometry
$M^{\bullet}$	radical species concentration
NBR	acrylonitrile–butadiene rubber
NMR	nuclear magnetic resonance
OM	optical microscopy
PA	polyacetylene
PAN	polyacrylonitrile
PAS	photoacoustic spectroscopy
PBT	poly(butylene terephthalate)
PC	polycarbonate
PCL	polycaprolactone
PCTFE	polychlorotrifluoroethylene
PDI	polydispersity index
PDMS	polydimethylsiloxane
PE	polyethylene
PEEK	poly(ether ether ketone)
PEES	poly(ether ether sulfone)
PEI	poly(ether imide)
PEO	poly(ethylene oxide)
PES	poly(ether sulfone)
PET	poly(ethylene terephthalate)
PGC	pyrolysis gas chromatography
PHEMA	poly(hydroxyethyl methacrylate)
PIP	piperidine
PMMA	poly(methyl methacrylate)
PP	polypropylene
PPO	poly(phenylene oxide)
PPP	poly(*p*-phenylene)

PPS	poly(phenylene sulfide)
PPY	polypyrrole
PS	polystyrene
PTFE	polytetrafluoroethylene
PU	polyurethane
PVA	poly(vinyl acetate)
PVAl	poly(vinyl alcohol)
PVC	poly(vinyl chloride)
$R^{\bullet}$	free radical
SALS	small-angle light scattering
SAN	styrene–acrylonitrile
SANS	small-angle neutron scattering
SAXS	small-angle X-ray scattering
SBR	styrene–butadiene rubber
SDS	sodium dodecyl sulfate
SEC	size-exclusion chromatography
SEM	scanning electron microscopy
SIMS	secondary-ion mass spectrometry
TCE	tetrachloroethane
TEM	transmission electron microscopy
TGA	thermogravimetric analysis
TLC	thin layer chromatography
TMA	thermal mechanical analysis
TMS	tetramethylsilane
TOF	time-of-flight
TPE	thermoplastic elastomer
UCST	upper critical solution temperature
U–F	urea–formaldehyde
UHMWPE	ultra-high-molecular-weight polyethylene
UTS	ultimate tensile strength
UV–Vis	ultraviolet–visible
WAXS	wide-angle X-ray scattering
WLF	Williams–Landel–Ferry
XPS	X-ray photoelectron spectroscopy

a	Mark–Houwink–Sakurada constant; capillary constant
a_T	shift factor
A	absorbance; area
B	field strength; osmotic virial coefficient
c	concentration
C_p	heat capacity (at constant pressure)
d	distance
d_{p}	depth of penetration

D	diffusion coefficient
E	Young's modulus
E'	storage modulus
E''	loss modulus
E^*	complex modulus
E_A	activation energy
E_B	binding energy
E_c	Young's modulus of composite
E_D	activation energy for transport of polymer chains
E_f	Young's modulus of fibre
E_K	kinetic energy
E_m	Young's modulus of matrix
f	initiator frequency; mole fraction in monomer feed
F	degree of orientation; flow rate; force; mole fraction of monomer
g	g-value
h	height; Planck constant
H	optical constant
H_0	external magnetic field strength
I	spin number; intensity
I_f	fluorescence intensity
I_p	phosphorescence intensity
I_0	incident intensity
J	coupling constant; gas compressibility factor
$J(t)$	creep compliance
k	rate constant; molar absorption coefficient
K	calibration factor; Mark–Houwink–Sakurada constant; proportionality constant
l	pathlength; bond length; length
m	mass
M	molecular weight
$\overline{M}_n$	number-average molecular weight
$\overline{M}_w$	weight-average molecular weight
m/z	mass-to-charge (ratio)
n	refractive index; Avrami exponent; melt flow index
$\overline{n}_n$	number-average degree of polymerization
N	number of molecules; number of bonds; cycles to failure
p	extent of reaction
P	pressure
$P(x)$	probability
q	heating rate; quantum yield
Q	rate of extrusion
r	radius; reactivity ratio

R	universal constant; X-ray distance; reaction rate
R_c	contour length
R_g	radius of gyration
R_{rms}	root-mean-square end-to-end distance
R_∞	absolute reflectance
$R(\theta)$	Rayleigh ratio
s	speed
S	sedimentation constant
S_0	ground singlet state
S_1	excited singlet state
t	time (general); elution time
T	transmittance; temperature
T_c	crystallization temperature
T_d	degradation temperature
T_g	glass transition temperature
T_m	melting temperature
T_1	excited triplet state
v	volume fraction; velocity
v_s	specific volume
V	volume
V_f	fibre volume
V_g	specific retention volume
V_m	matrix volume
V_R	retention volume
w	weight
w_i	weight fraction
x	X-ray line spacing
x_c	mass fraction of crystals
x_i	number fraction
x_n	number-average chain length
x_w	weight-average chain length
Z_c	critical chain length
Z_w	weight-average chain length
α	coefficient of thermal expansion
β	Bohr magneton
γ	magnetogyric ratio; shear rate
γ_c	critical surface tension
γ_L	surface tension
δ	chemical shift; phase angle; solubility parameter
ΔC_p	heat capacity change
ΔE	energy change
ΔG_m	Gibbs free energy change on mixing

ΔH	enthalpy change
ΔH_a	enthalpy change of amorphous standard
ΔH_c	enthalpy change of crystalline standard
ΔH_m	enthalpy change on mixing
Δn	birefringence
ΔS_m	entropy change on mixing
ΔT	temperature change
ε	molar absorptivity; strain
η	viscosity
$[\eta]$	intrinsic viscosity
θ	angle of incidence; contact angle; Flory temperature
θ_d	dynamic contact angle
θ_s	static contact angle
λ	wavelength; extension ratio
μ	friction coefficient
ν	frequency; linear spherulite growth rate
ν_0	universal constant for crystalline polymers
π	osmotic pressure
ρ	density
ρ_a	density of amorphous polymer
ρ_c	composite density; density of crystalline polymer
ρ_f	fibre density
ρ_m	matrix density
ρ_s	density of polymer sample
σ	stress
σ_c	tensile strength of composite
σ_f	tensile strength of fibre
σ_m	tensile strength of matrix
σ_Y	yield strength
τ	shear stress; relaxation time
ϕ	angle of refraction
Φ	work function
ω	angular velocity

About the Author

Barbara Stuart, B.Sc., M.Sc. (Hons), Ph.D., DIC, MRACI, CChem MRSC

After graduating with a B.Sc. degree from the University of Sydney in Australia, Barbara Stuart then worked as a tutor at this university. She also carried out research in the field of biophysical chemistry in the Department of Physical Chemistry and graduated with an M.Sc. degree in 1990. The author then moved to the UK to carry out doctoral studies in polymer engineering within the Department of Chemical Engineering and Chemical Technology at Imperial College (University of London). After obtaining her Ph.D. in 1993, she took up a position as a Lecturer in Physical Chemistry at the University of Greenwich in South East London. Barbara returned to Australia in 1995, joining the staff of the Department of Materials Science at the University of Technology, Sydney, where she is currently a Senior Lecturer. She is presently conducting research in the fields of polymer spectroscopy, materials conservation and forensic science. Barbara is the author of two other books published by John Wiley and Sons, Ltd, namely *Modern Infrared Spectroscopy* and *Biological Applications of Infrared Spectroscopy*, both of these titles in the ACOL Series of open learning texts.

Chapter 1

Introduction

Learning Objectives

- To understand the basic definitions used to describe polymers.
- To understand the different categories of polymers as based on their structures.
- To appreciate the history of the development of synthetic polymers.
- To recognize common thermoplastics, thermosets and elastomers, and their specific properties.
- To recognize the structures and properties of high-performance polymers.
- To understand the nature and types of copolymers.
- To understand the characteristics of polymer blends.
- To understand the nature and composition of polymer composites.
- To recognize the types of additives used in polymers.
- To understand the nature of speciality polymeric materials, such as liquid crystalline polymers, conducting polymers, thermoplastic elastomers, biomedical polymers and biodegradable polymers.

1.1 Introduction

Polymers play an enormously important role in modern society. The significance of these materials is often taken for granted, yet polymers are fundamental to most aspects of modern life such as building, communication, transportation, clothing and packaging. Thus, an understanding of the structures and properties of polymeric materials is vital.

What is a polymer? Polymers are large molecules consisting of a large number of small component molecules. In fact, the name polymer derives from the

Greek 'polys' meaning 'many' and 'meros' meaning 'part.' Many polymers are synthesized from their constituent *monomers* via a *polymerization* process. Most commercial polymers are based on covalent compounds of carbon, although certain synthetic polymers may also be based on inorganic atoms such as silicon.

Vinyl polymers have names which are derived from the names of their particular monomers. For example, poly(vinyl chloride) (PVC) is made from vinyl chloride ($CH_2{=}CHCl$). PVC is usually denoted as follows:

$$\left[CH_2 - \underset{\displaystyle Cl}{\underset{|}{CH}} \right]_n$$

illustrating the structural repeat unit of the polymer. The repeat unit in a polymer is often referred to as a *mer*. The n in the polymer structure is known as the *degree of polymerization* (DP) and refers to the number of 'mers' in a polymer structure.

SAQ 1.1

What is the degree of polymerization of a sample of polyethylene, [$-(CH_2{-}CH_2)_n-$], which has a molecular weight of 100 000 g mol^{-1} ?

Polymers can display a range of different structures (see Figure 1.1). In the simplest case, they possess a simple *linear* structure. However, polymers can also be *branched*, depending on the method of polymerization. They may also display a *cross-linked* structure. Some more unusual polymer structures include *star* polymers, which contain three or more polymer chains connected to a central unit, *ladder* polymers, which consist of repeating ring structures, and *dendrimers*, which show a star-like structure with branching. These different sorts of microstructures have an effect on the properties of the polymer; this aspect will be discussed further in Chapter 5.

Polymers are often commonly referred to as 'plastics'. However, this is somewhat of a misnomer. The term 'plastic' refers to one class of polymers known as *thermoplastics*. Polymers in this category show a range of different properties, but a simple definition is to describe these as polymers that melt when heated and re-solidify when cooled. Thermoplastics tend to be made up of linear or lightly branched molecules, as such structures enable the polymer chains enough freedom of movement to change form as a function of temperature. However, not all polymers are capable of being melted. For example, *thermosets* are polymers that do not melt when heated, but decompose irreversibly at high temperatures. Thermosets are cross-linked, with the restrictive structure preventing melting behaviour. Some cross-linked polymers may show rubber-like characteristics and these are known as *elastomers*. Such materials can be extensively stretched but will rapidly recover their original dimensions.

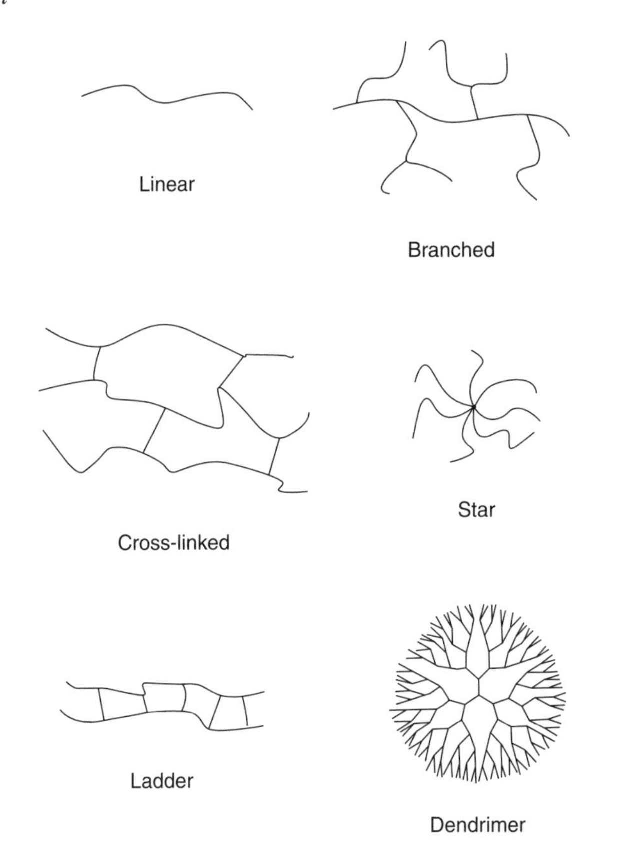

Figure 1.1 Various types of polymer structures.

1.2 History

The modern polymer industry evolved from the modification of the properties of certain natural polymers [1]. In the 19th century, the latex extracted from a tropical rubber tree was used to produce a rubbery material. The US chemist, Charles Goodyear, was able to improve the elastic properties of natural rubber by heating with sulfur, a process known as *vulcanization.*

Cellulose nitrate, derived from cellulose, was developed by Christian Schoenbein in Switzerland during 1846. Although initially recognized as an explosive, it was soon realized that cellulose nitrate was also a hard elastic material which could be readily moulded into different shapes. The development of *celluloid*, a plasticized version of cellulose nitrate, soon followed and was a commercial success, leading to the development of photography. By the late 19th century, other modifications of cellulose, such as viscose rayon fibres and cellophane, had been developed.

The first decade of the 20th century saw the development of the first synthetic plastics. In the United States, Leo Baekeland reacted phenol and formaldehyde to obtain a heat-resistant material that was marketed as 'Bakelite'. This polymer soon achieved broad commercial success and became widely used for household goods, and in the developing electronic and car industries. Despite the success of Bakelite, one disadvantage was its dark colour. Consequently, in the 1920s, urea–formaldehyde polymers were introduced, with such materials being available in a wide range of colours.

The 1930s saw the development of the commercially important material, *polyethylene*. Chemists at ICI in the UK, while experimenting with ethylene at different temperatures and pressures, stumbled upon this important polymer. The development of *nylon* was a more deliberate process. Wallace Carothers of the DuPont Company in the United States set about producing a material which could replace silk. The 2nd World War was responsible for the development of many synthetic polymers, with war-time needs forcing the production of low-cost plastics. By the 1950s, polyethylene, polystyrene and poly(vinyl chloride) had become widely available materials and very much a part of everyday life. Since this period, polymers have been developed and incorporated into many aspects of life – everything from engineering to medicine.

1.3 Thermoplastics

Thermoplastics are polymers that require heat to make them processable. After cooling, such materials retain their shape. In addition, these polymers may be reheated and reformed, often without significant changes to their properties. Many thermoplastics contain long main chains consisting of covalently bonded carbon atoms. Table 1.1 summarizes the properties and applications of some common thermoplastics [2].

Polyethylene (PE) is the major general purpose thermoplastic and is widely used for packaging, containers, tubing and household goods. The structural repeat unit of PE is as follows:

$$\left[CH_2 - CH_2 \right]$$

The main reasons for the popularity of PE is its low cost, easy processability and good mechanical properties. A susceptibility to weathering is a limitation, but this is not usually a problem with routine applications of PE. There are two main types of mass-produced polyethylene. *Low-density polyethylene* (LDPE) has a branched chain structure and tends to be used for bags and packaging, while *high-density polyethylene* (HDPE) has a mostly linear structure and finds uses in bottles and containers. *Linear low-density polyethylene* (LLDPE) has also been developed for its good processing properties. This PE has a linear chain structure with short side branches and is used for bags. Another class of PE of note is

Table 1.1 The properties and applications of some common thermoplastics

Name	Trade names	Properties	Applications
Polyethylene (PE)	Polythene, Rigidex, Alkathene, Hostalen, Lupolen, Alathon	Inexpensive, easily processed, good chemical resistance, poor resistance to weathering	Household goods, packaging, containers
Polypropylene (PP)	Propathene, Appryl, Novolen	Inexpensive, good chemical resistance, poor ultraviolet resistance	Packaging, containers, furniture, pipes
Poly(vinyl chloride) (PVC)	Darvic, Corvic, Geon, Evipol, Vinnolit, Hostalit	Inexpensive, rigid, good chemical resistance, limited thermal stability, additives required for processing	Packaging, cable insulation, pipes, toys
Polystyrene (PS)	Styron, Polystyrol, Novacor	Inexpensive, transparent, rigid, good insulating properties, low water absorption, flammable, brittle	Food containers, packaging, appliance housings
Poly(methyl methacrylate) (PMMA)	Perspex, Plexiglas, Lucite, Acrylite	Transparent, good weathering properties, tough, rigid, poor insulating properties, poor resistance to organic solvents	Transparent sheets and mouldings, aeroplane windows, street lamps, display signs
Polyamide (PA)	Nylon, Ultramid, Zytel, Caprolan, Stanyl, Capron, Akulon, Rilsan, Vestamid	Tough, flexible, abrasion resistant, good wear and frictional properties, absorbs water	Textiles, brushes, surgical applications, bearings, gears
Polytetrafluoroethylene (PTFE)	Teflon, Fluon, Halon, Hostaflon	Low friction, good electrical insulation, excellent chemical resistance, cannot be dissolved, relatively expensive	Non-stick surfaces, insulation tape, engineering applications
Polyacrylonitrile (PAN)	Barex, Orlon	Strong, good chemical resistance	Wool-type applications
Cellulose acetate	Tenite, Acetate, Clarifoil, Dexel	Crease resistance, moisture resistance, dyeability	Textile fibres, moulded products, film, packaging
Poly(vinyl acetate) (PVA)	Elvacet, Vinylite	Good general stability, quick drying, inexpensive	Surface coatings, adhesives, paint
Poly(vinyl alcohol) (PVAl)	Vinex	Water-soluble	Fibres, adhesives, thickening agents
Poly(ethylene terephthalate) (PET)	Terylene, Dacron, Melinex, Mylar	Low short-term water absorption, fibres are crease resistant, strong, high processing temperatures required	Textile fibres, film, packaging, magnetic tapes
Poly(butylene terephthalate) (PBT)	Celanex, Tenite, Rynite, Valox	Strong, good chemical resistance, good electrical insulation	Electrical, electronic and automative engineering

ultra-high-molecular-weight polyethylene (UHMWPE), a linear PE with a high molecular weight of the order of 4×10^6 g mol^{-1}. This form of polyethylene shows excellent wear and abrasion resistance, high impact resistance and a very low friction coefficient. UHMWPE also possesses good chemical resistance and a self-lubricating and non-stick surface. This combination of properties leads to its use in diverse applications such as medical prostheses, blood filters, bullet-proof vests and fishing lines.

Polypropylene (PP) shows a similar structure to PE, but with a substituted methyl group, as follows:

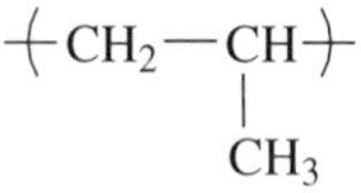

$$\left[CH_2 - \underset{\displaystyle CH_3}{\underset{|}{CH}} \right]$$

The presence of the methyl group restricts the rotation of the PP chain and produces a less flexible, but stronger polymer. Like polyethylene, polypropylene shows several attractive properties such as good chemical and moisture resistance and high dimensional stability. These characteristics make this polymer suitable for a wide range of applications, such as bottles, carpets, casings and packaging.

Poly(vinyl chloride) (PVC) is the second largest volume thermoplastic polymer. PVC has a structure which contains a chlorine atom on alternate main chain carbons, as follows:

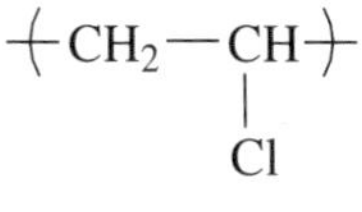

$$\left[CH_2 - \underset{\displaystyle Cl}{\underset{|}{CH}} \right]$$

There are strong dipole interactions between the PVC chains and steric hindrance reduces the flexibility of the molecules. This lack of flexibility in PVC molecules means that this polymer can only be processed when compounded with various *plasticizers*. PVC is widely used because of its excellent chemical resistance and its ability to be modified with additives. In its rigid form, PVC is commonly used in construction as piping, guttering, siding and electrical conduit. Plasticized PVC is also used widely as electrical wire insulation and in household and automotive applications.

Polystyrene (PS) is another widely used thermoplastic, being the fourth most produced polymer by weight. PS is a clear, rigid and brittle material, unless it is modified with rubber. The polymer molecule contains a benzene ring attached to alternate carbon atoms on the backbone, as follows:

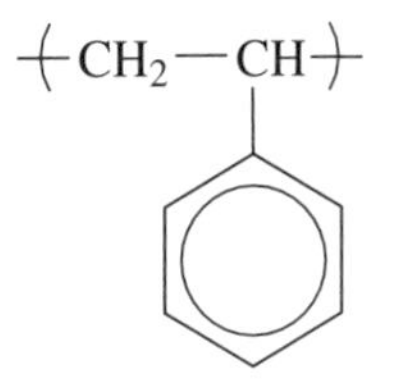

The presence of this aromatic group causes rigidity in the structure due to steric hindrance. The popularity of this polymer results from its low cost, low water absorption and good insulating properties. PS is used widely in packaging and appliance housings.

Poly(methyl methacrylate) (PMMA) is the best known type of *acrylic* thermoplastic, having the following repeat unit:

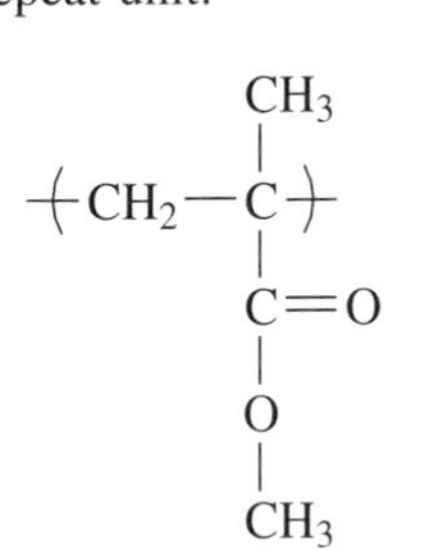

Well known by the trade name 'Perspex', this polymer is a transparent, hard and rigid material. The steric hindrance resulting from the large side group in the structure is responsible for the rigidity. The good outdoor weatherability and very good impact resistance of PMMA means that this polymer is particularly useful in glazing, street furniture and skylights.

Polyamides are probably better known as *nylons*. There are a number of different types of nylons, although all of these contain an amide linkage. Single-number nylons are so named because of the number of carbon atoms contained in the structural repeat unit. For example, nylon 6 has the following structure:

$$\left[NH-(CH_2)_5-CO \right]_n$$

with six carbon atoms contained in its repeat unit. Nylon 6 is a common polyamide, while nylon 11 and nylon 12 are two other well established single-number nylons. Double-number nylons are named by counting the number of carbon atoms in the amide section and the number in the carbonyl section of the repeat unit. For example, the main double-number nylon is nylon 6,6, which contains the following structural repeat unit:

$$\left(NH-(CH_2)_6-NH-CO-(CH_2)_4-CO \right)$$

Nylon 6,10 and nylon 6,12 are two other well established double-number nylon materials. Nylons, in general, are engineering thermoplastics with good wear and frictional properties and toughness. Such properties are, in part, due to the hydrogen bonding between the nylon molecule chains. Nylons are employed in a wide range of uses, including textile, engineering, electrical and electronic applications.

Polytetrafluoroethylene (PTFE) is perhaps better known by its trade name of 'Teflon'. There is a common misconception that 'the only good thing that came out of the space race was the non-stick frypan', referring to the emergence of PTFE as a common surface coating material in the 1960s. In fact, PTFE was originally discovered in 1938 by Roy Plunkett at DuPont in the United States when he accidentally polymerized tetrafluoroethylene gas. PTFE is an engineering thermoplastic which shows remarkable chemical resistance, electrical insulating properties and a low friction coefficient. However, it does have the disadvantage that it is difficult to process on account of its high melting temperature. PTFE is a dense crystalline polymer with the following structural repeat unit:

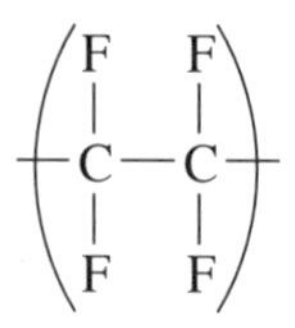

PTFE is used for non-stick coatings, electrical components, bearings and tapes. There are several other halogen-containing thermoplastics, including *polychlorotrifluoroethylene* (PCTFE), $[-(CF_2-CFCl)-_n]$, and poly(vinylidene chloride) (PVDC), $[-CH_2-CCl_{2n}-]$.

Polyacrylonitrile (PAN) is commonly used as a fibre because of its chemical resistance and its high strength. PAN is also used in various copolymer formulations. The structural repeat unit of PAN is as follows:

$$-\!\left(CH_2-\underset{\underset{\displaystyle C\equiv N}{|}}{CH}\right)\!-$$

The PAN molecules form extended structures due to the high electronegativity of the nitrile group. The resulting hydrogen bonding between the polymer chains allows strong fibres to be produced. PAN fibres are used in wool-type applications, such as blankets.

Cellulose, the major component of cotton, is a naturally occurring polysaccharide, with the following structure:

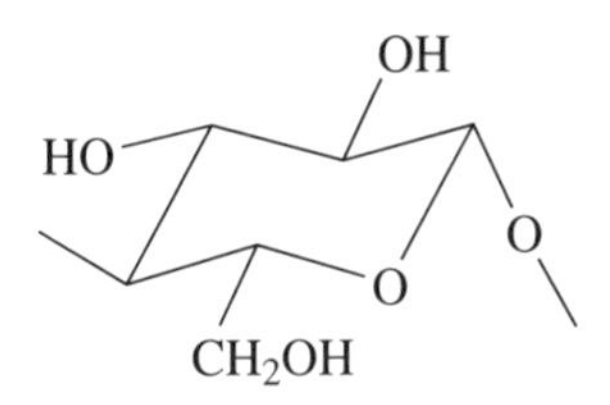

When cellulose is dissolved via a particular chemical reaction and then reprecipitated as pure cellulose, the product is known as *regenerated cellulose*. When the latter is prepared as a fibre, it is known as *viscose* or *viscose rayon* and has been widely used for textile fibres. When prepared as a film, regenerated cellulose is known as *cellophane*, a well-known packaging and wrapping material. There are several different chemical derivatives of cellulose. *Cellulose nitrate* (or nitrocellulose) was an early synthetic plastic, being first discovered in 1838. This derivative was plasticized with camphor and manufactured as 'Celluloid', which led to the development of the cinema industry. Cellulose nitrate has been superseded in this type of application because of its flammability and degradability, but now finds use in the field of coatings as a lacquer. The discovery of the more commercially important *cellulose acetate* followed in 1865 as a result of the esterification of cellulose. When complete acetylation is carried out, *cellulose triacetate* is formed, as follows:

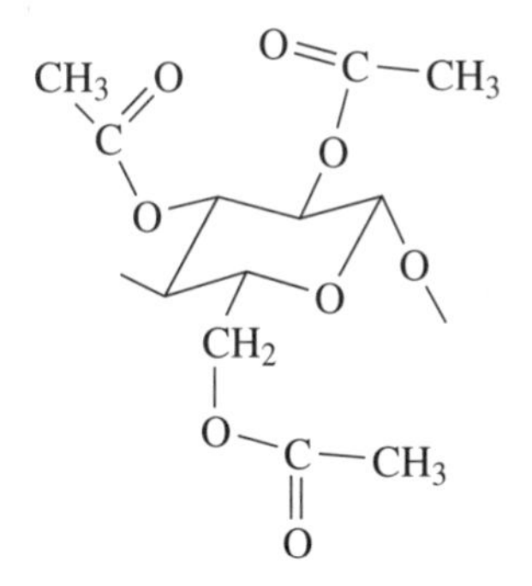

However, the acetylation reaction may also be reversed to a point where *cellulose diacetate* is formed. This diacetate is commonly referred to as *cellulose acetate* and is more suitable for use as a fibre. The triacetate derivative is fully esterified and is known as the *primary acetate*. Note that there are not necessarily three acetate groups per repeat unit. The degree of substitution can be averaged to, for example, 2.8. The diacetate, with a degree of substitution of about 2, is called the *secondary acetate*. By varying the reaction conditions, different acetate contents can be obtained, thus yielding products with properties suitable for specific purposes. Cellulose acetates are employed in a wide range of forms, including fibres, moulded products, films and packaging.

Poly(vinyl acetate) (PVA) is used as a thermoplastic, mostly in the form of an emulsion, and also as a precursor for the preparation of *poly(vinyl alcohol)* (PVAl). PVA is tough and stable at room temperature, but becomes sticky and flows at slightly elevated temperatures. PVA has the following structural repeat unit:

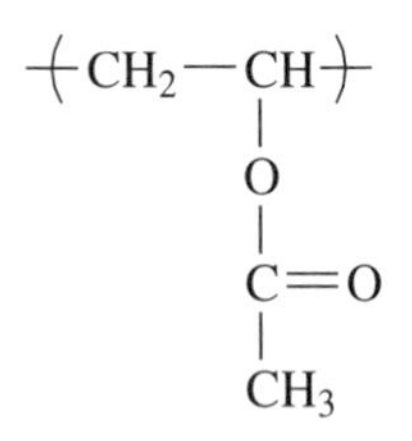

PVA is a stable, low-cost and quick-drying polymer and so can be used in the production of water-based paints. It is also used commonly as an adhesive. Alcohols are added to PVA to produce PVAl, which is a water-soluble polymer used for fibres, adhesives and as thickening agents. The repeat unit of PVAl is as follows:

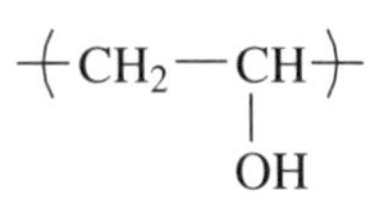

There are two important thermoplastic polyesters which find wide use. *Poly(ethylene terephthalate)* (PET) is routinely used for textile fibres and food packaging, but is also used as a container resin. PET shows good short-term water absorption and the fibres are strong and crease-resistant. The structural repeat unit of PET is as follows:

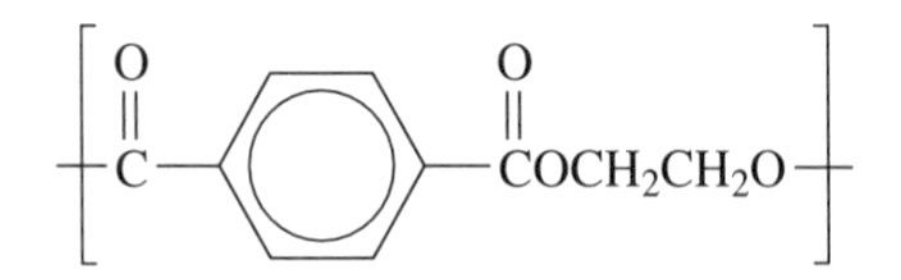

Poly(butylene terephthalate) (PBT) has found expanding applications in the electrical, electronic and automotive industries. The structural repeat unit of PBT is as follows:

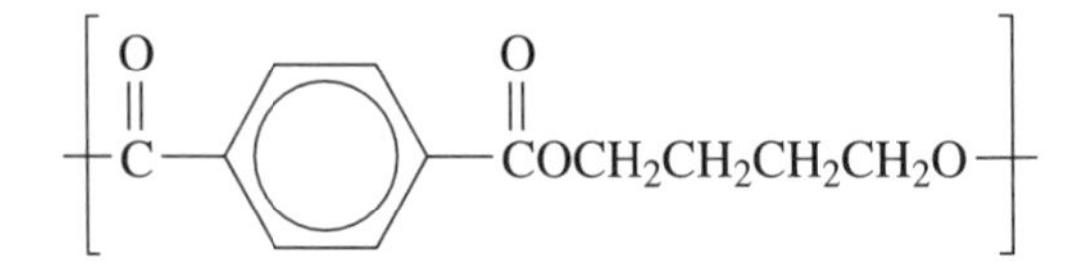

The benzene rings in the PBT structure provide rigidity, while the butylene units provide a degree of molecular mobility, thus enabling melt processing. PBT is a strong resin, resistant to most chemicals and with good electrical insulation properties.

DQ 1.1

What would be a suitable thermoplastic for use as glazing for a bus shelter?

Answer

PMMA is suitable for this sort of outdoor application as it is transparent, rigid and has good weatherability.

1.4 Thermosets

Thermosets possess a networked (cross-linked) structure. Such a structure may be formed by heating or via a chemical reaction. Thermosets tend to possess excellent thermal stability and rigidity. Table 1.2 summarizes the properties and applications of some common thermosets [2].

Phenol–formaldehyde (or *phenolic*) resins were the first major industrial polymers, although they remain in wide use today because they are low in cost and possess good electrical, insulating and mechanical properties. Phenolic resins are

Table 1.2 The properties and applications of some common thermosets

Name	Trade names	Properties	Applications
Phenol–formaldehyde resins	Bakelite, Plenco, Durite	Good chemical, thermal and flame resistance	Electrical mouldings, appliance handles, household fittings, adhesives, laminates
Epoxy resins	Araldite, Epon, Epikote	Tough, good adhesion, excellent chemical resistance, good electrical properties	Adhesives, protective coatings, laminates, building and construction, electrical and electronic components
Amino resins	Avisco, Plaskon	Rigid, strong, good impact resistance	Kitchenware, flooring, particleboard, plywood
Polyester resins	Laminac, Aropol, Baygal	Low viscosity, inexpensive	Automative engineering, construction, boat building
Polyurethanes	Lycra, Elastane	Burn easily producing toxic fumes, good abrasion and chemical resistance, very high elasticity, excellent abrasion, adhesion and impact properties	Furniture, tyre treads, clothing, floor and kitchen surfaces

produced by reacting phenol and formaldehyde. The cross-linking reaction in phenolic resins is carried out on prepolymers that have been formed by having one of the components in excess in order to minimize cross-linking during the initial reaction step. There are two main types of prepolymers, namely *novolaks* and *resoles*. Novolaks are formed with phenol in excess and under acidic conditions and show structures such as the following:

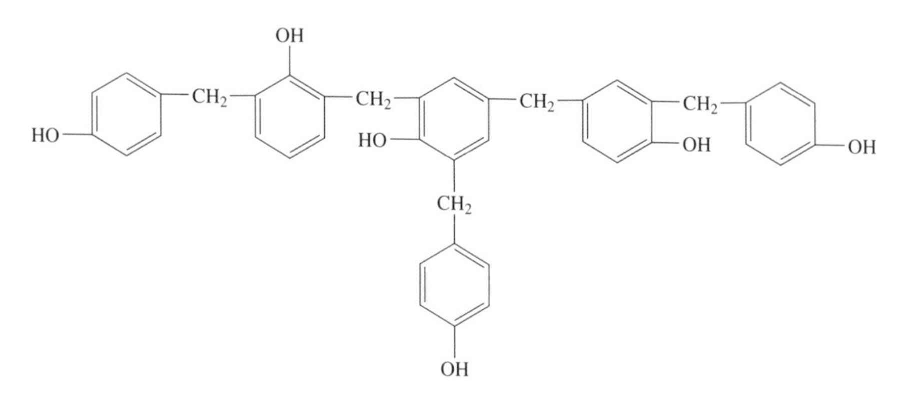

Resoles are formed with an excess of formaldehyde and under basic conditions and show the following type of structure:

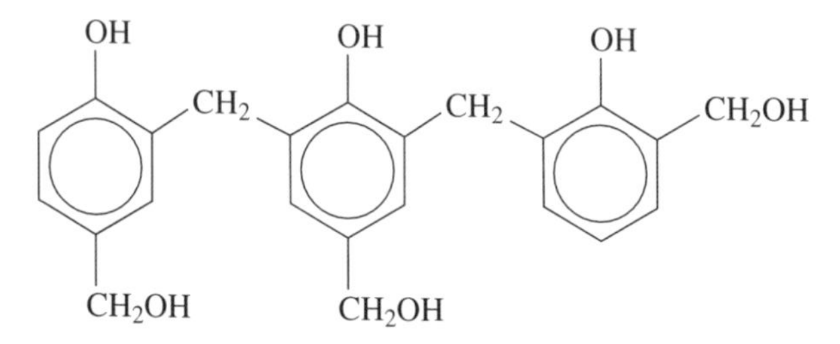

Phenolic resins are used for electrical equipment, automotive components and in household appliances such as handles (novolaks). They also find use as high-temperature adhesives and laminates (resoles).

Epoxy resins are thermosetting resins that contain epoxide groups. The most common type of epoxy prepolymer is based on glycidyl ethers. For example, the epoxy diglycidyl ether of bisphenol A shows the following structure:

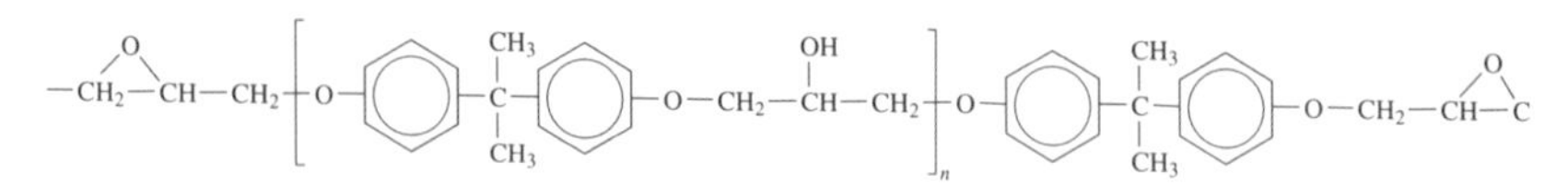

The resins are cured by using catalysts or cross-linking agents such as amines and anhydrides. The epoxy and hydroxyl groups are the reaction sites for cross-linking and can undergo cross-linking reactions that result in no by-products. This results in low shrinkage during hardening. Epoxy resins show good adhesion and mechanical properties and good chemical resistance, as well as being good electrical insulators. Consequently, they are used in coatings, composites and electrical and electronic components.

Polyurethanes (PUs) are versatile thermosets that are used as foams, elastomers, fibres, adhesives and coatings. PUs display a range of chemical compositions, although they all contain the common urethane group, as follows:

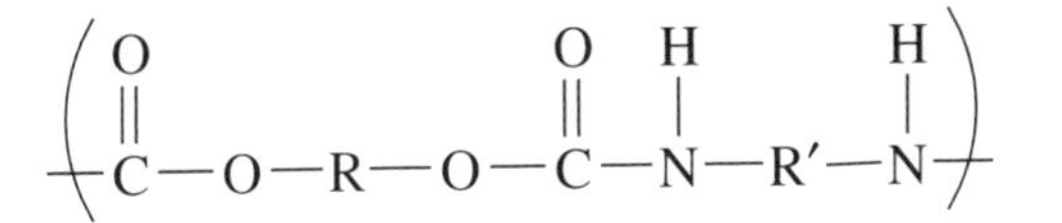

Urethanes are formed by the reaction between a hydroxyl-containing molecule and a reactant containing an isocyanate group. If this reaction is extended by using compounds containing two or more reactive groups, a complex polymeric structure is produced. The reaction of a diisocyanate and a polyhydroxyl compound (a *polyol*) will produce a cross-linked polymer structure. In the production of urethane foam, an excess of isocyanate groups in the polymer react with water to produce carbon dioxide, thus 'blowing' the foam at the same time that cross-linking occurs. Such foams may be rigid or flexible, depending upon the polymer type and the degree of cross-linking.

PU elastomers are made by first preparing a basic polyester or polyether intermediate in the form of a low-molecular-weight polyol. This intermediate is then reacted with a diisocyanate to a give a prepolymer. The elastomer is then vulcanized via the isocyanate groups. PU elastomers show good abrasion and chemical resistance, and are used in tyre treads. PU fibres, such as 'Lycra', exhibit unusually high elasticity and are now widely used for lightweight clothing. PU coatings show excellent abrasion, adhesion and impact properties and have found expanding use as floor and kitchen surface materials.

Formaldehyde can be reacted with various amine-containing groups to form thermosetting *amino resins*. *Urea–formaldehyde resins* (U–F resins) are formed by a condensation reaction between urea and formaldehyde. The amine groups at the end of the urea–formaldehyde molecule react with more formaldehyde molecules to form the network structure shown in Figure 1.2(a). The urea and formaldehyde are first partially polymerized to form a low-molecular-weight polymer. This can then be ground into a powder, compounded with fillers such as cellulose, and then moulded into the required shape.

Melamine–formaldehyde resins (M–F resins) are formed by a similiar type of condensation reaction, with the type of structure produced illustrated in Figure 1.2(b).

Amino resins are rigid, strong and show good impact resistance. M–F resins have better heat resistance than U–F resins, although the latter resins tend to be less expensive to produce. Amino resins are used for kitchenware, flooring, particleboard and plywood.

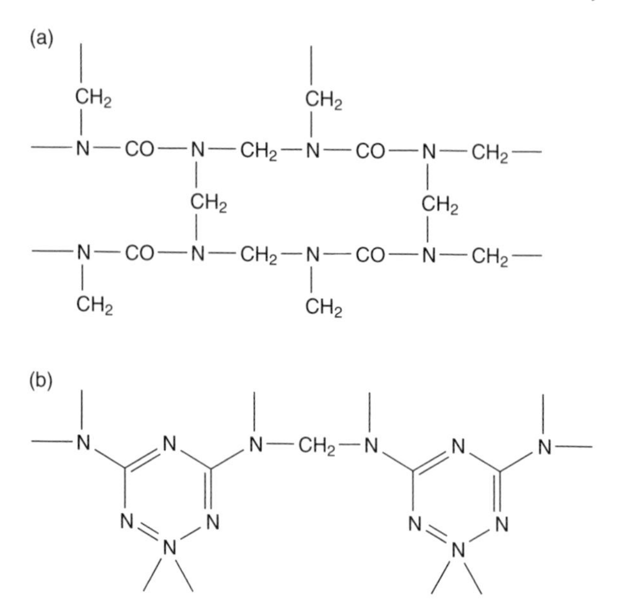

Figure 1.2 Structures of two important types of amino resins: (a) urea–formaldehyde; (b) melamine–formaldehyde.

Unsaturated polyesters can be cross-linked to form thermosets and are commonly used with glass fibres to produce high-strength composites. Linear polyesters are cross-linked with vinyl monomers, such as styrene, in the presence of a free-radical curing agent, as follows:

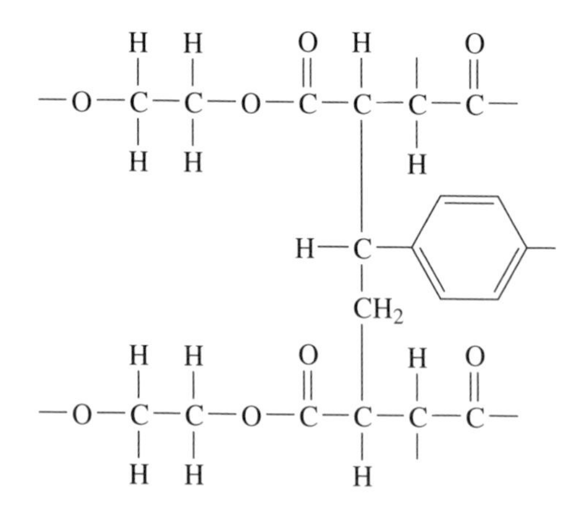

As *polyester resins* are low in viscosity, they can readily be mixed with glass fibres. As a consequence, they are widely used in construction, automotive engineering and boat building.

DQ 1.2

Suggest a suitable thermoset for use as a kitchen-bench coating.

Answer

Although a number of resins have been used for this type of application, PUs are currently popular for use as coatings for domestic purposes.

1.5 Elastomers

Elastomers, or rubbers, are polymers that may have their dimensions changed when stressed. When the stress is removed, the polymer returns to its original dimensions. Table 1.3 summarizes the properties and applications of some common elastomers [2].

The first recognized elastomer was *natural rubber*. Such a rubber is extracted from the latex of the tropical *Hevea brasiliensis* tree. This material consists mainly of poly(*cis*-isoprene) mixed with small amounts of other components, including proteins and lipids. The structural repeat unit of poly(*cis*-isoprene) is as follows:

$$\begin{array}{ccc} CH_3 & & H \\ \diagdown & & \diagup \\ & C{=}C & \\ \diagup & & \diagdown \\ -\!\!\left(CH_2\right. & & \left.CH_2\right)\!\!- \end{array}$$

Table 1.3 The properties and applications of some common elastomers

Name	Trade names	Properties	Applications
Poly(*cis*-isoprene) (natural rubber)	Natsyn, Amenpol	Good abrasion resistance, poor heat and oil resistance, good electrical properties	Gaskets, shoe soles, condoms
Silicone	Silastic, Rimplast, Silane	Thermally stable, water resistant, expensive	Medical implants, sealants, flexible moulds, gaskets, electrical insulation
Polychloroprene	Neoprene, Perbunan	Oil resistance, good weathering, low flammability	Cables, hoses, seals, gaskets

Note that the poly(*trans*-isoprene) isomer – known as *gutta percha* – is a hard brittle material. In order for the extracted natural rubber to form an elastomer, a cross-linking process known as *vulcanization* must be carried out. Vulcanization is the process by which elastomers are lightly cross-linked in order to reduce plasticity and to develop elasticity. Natural rubber is not elastic until it has been lightly cross-linked. When natural rubber is heated with sulfur, the latter forms cross-links between the polyisoprene chains, as illustrated in Figure 1.3. Natural rubber shows good abrasion resistance and electrical properties and is used for applications ranging from shoe soles to contraceptives.

SAQ 1.2

How much sulfur must be added to 100 g of polyisoprene rubber to cross-link 5% of the monomer units? Assume that all of the sulfur is used and that only one sulfur atom is involved in each cross-link.

The structures of *silicone rubbers* are based on silicon and oxygen, with the structural repeat unit given by the following:

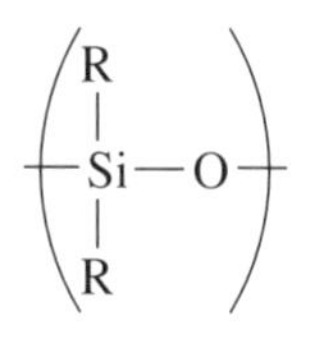

where R can be a hydrogen atom or groups such as methyl or phenyl. The most common silicone elastomer is polydimethylsiloxane (PDMS), which may be cross-linked to form Si–CH_2–CH_2–Si bridges. These elastomers are thermally stable and water resistant, but can be relatively expensive. Silicone rubbers are used for medical applications, sealants, gaskets and in electrical insulation.

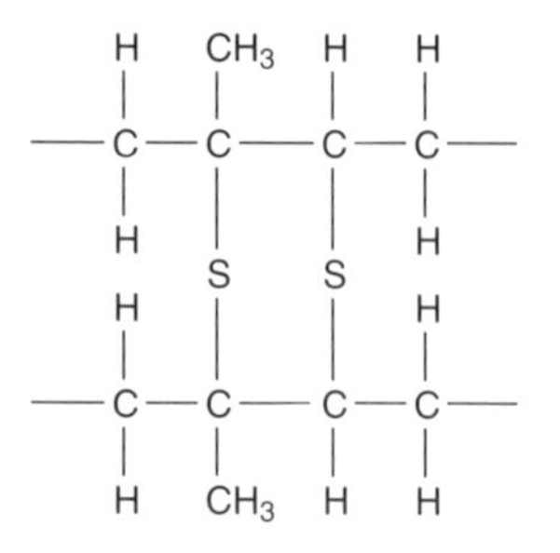

Figure 1.3 Illustration of the vulcanization process of natural rubber, in which sulfur forms cross-links between the polyisoprene chains.

Polychloroprene, commonly known as neoprene, has the following structural repeat unit:

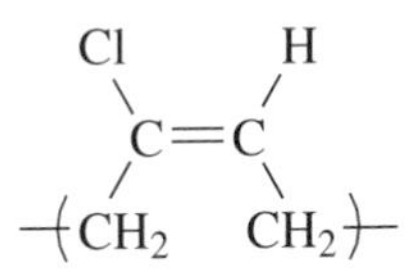

This elastomer shows low flammability, plus better weathering properties and oil resistance than other synthetic elastomers, although higher in cost. Polychloroprene tends to be used mainly for cable coverings, hoses and seals.

DQ 1.3

What would be suitable elastomer for use as a heart valve?

Answer

Silicone elastomers are used to manufacture heart valves as silicone-based elastomers show good biocompatability. PU elastomers can also be used as their chemistry can be tailored specifically to produce a biocompatable material. The biomedical applications of polymers are discussed further in Section 1.11 below.

1.6 High-Performance Polymers

Polymers that can cope with stringent high-temperature engineering environments are termed *high-performance polymers* (HPPs) [3]. Many of the established thermoplastics are not capable of withstanding high-temperature environments, so more recent work has focussed on polymer structures containing, for instance, thermally stable aromatic groups and resonance-stabilized systems. HPPs show the common characteristics of toughness, hardness, rigidity, high-temperature resistance, low flammability and low smoke emission. This class of polymers is found in automotive, aerospace, electrical, electronic and industrial applications.

Polycarbonate (PC) (trade names include 'Lexan', 'Makrolon' and 'Merlon') is a tough, strong and dimensionally stable engineering thermoplastic. The structural repeat unit of bisphenol A polycarbonate is as follows:

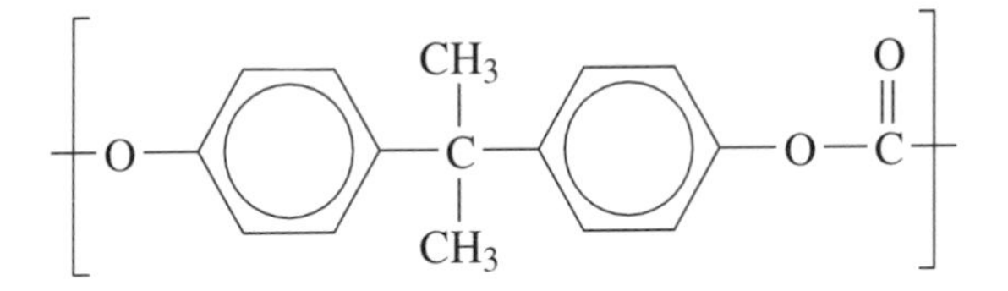

While the phenyl and methyl groups produce steric hindrance and, thus, a stiff molecular unit, the C–O bonds provide some flexibility in the structure. PC is used for compact discs, glazing, computer housings and aeronautical engineering applications.

Poly(phenylene oxide) (PPO)-based resins (trade names include 'Noryl', 'Vesteron', and 'Luranyl') are strong, rigid, dimensionally stable and chemically resistant HPPs. The structural repeat unit of PPO is as follows:

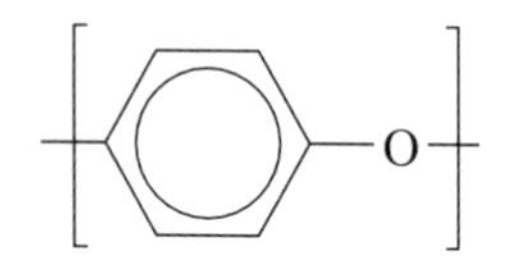

Steric hindrance to rotation is introduced by the benzene ring and there is electronic attraction due to the resonating electrons in the aromatic rings of adjacent molecules. PPO resins have found use in electrical and automotive engineering.

Polysulfones (trade names include 'Udel', and 'Ultrason') are strong, tough and transparent HPPs, with the following structural repeat unit:

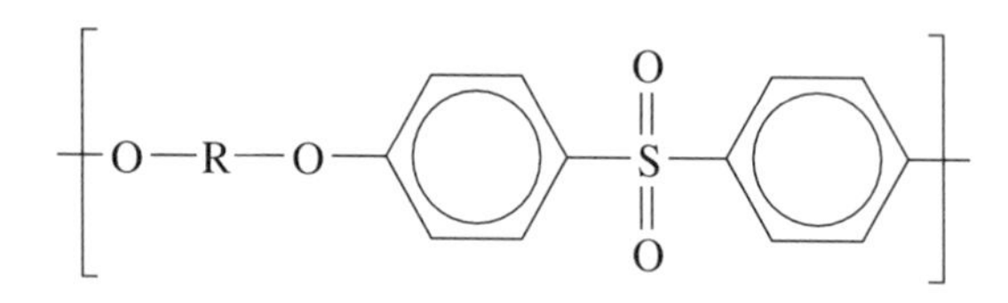

The benzene rings restrict rotation and provide for a strong intermolecular attraction. The oxygen atoms of the phenylene ring provide for a high oxidation stability, while the oxygen atoms in the ether linkage provide chain flexibility. Polysulfones have found use in electronic and electrical applications, and also in medical equipment because of their ability to be autoclave-sterilized.

One of the best groups of HPPs showing exceptional temperature performance is the *polyimides* (trade names include 'Ultem', 'Kapton' and 'Vespel'). Polyimides are produced using aromatic dianhydrides and aromatic amines to give the following general structure:

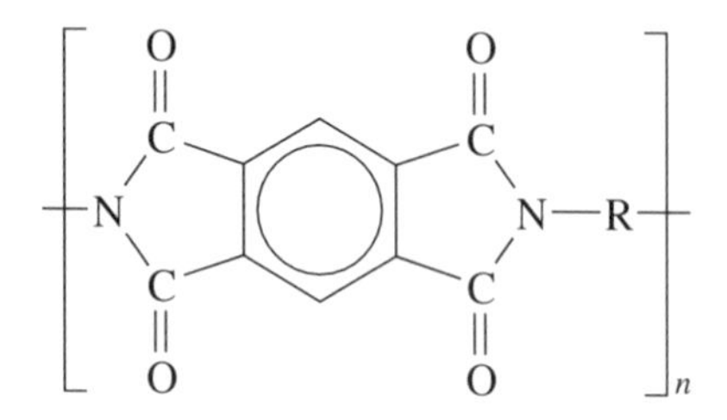

Commercial polyimides often contain ether units in the structure in order to aid processability. An important polyimide is *poly(ether imide)* (PEI), which has the following structure:

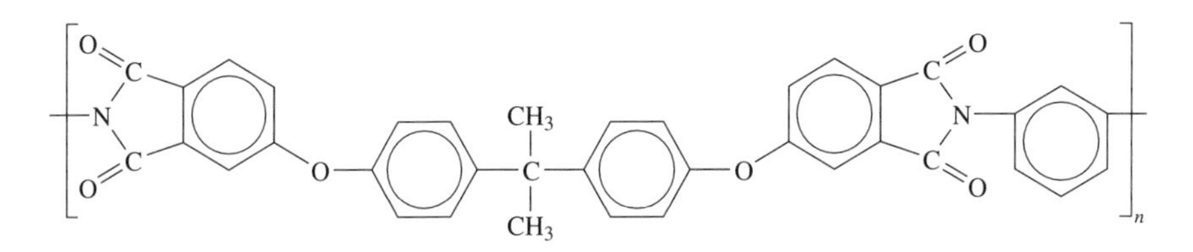

PEI is a rigid material due to the stable imide group, while the ether linkage between the benzene groups provides a degree of chain flexibility, thus allowing for melt processability. PEI also possesses good electrical insulation properties and, consequently, has found use in electrical and electronic applications. The polymer is also very suitable for automotive and aerospace applications.

Poly(phenylene sulfide) (PPS) (trade names include 'Ryton' and 'Fortron') possesses the following structural repeat unit:

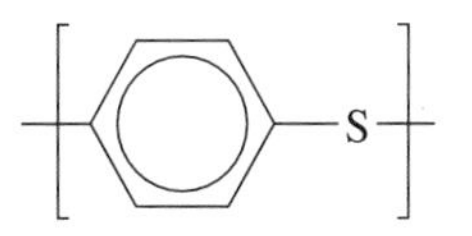

PPS is strong and rigid due to its symmetrical and compact structure. This polymer is also highly chemically resistant, due to the presence of the sulfur atoms. This property makes PPS suitable for use in industrial or mechanical environments such as in chemical process equipment, and in automotive applications. PPS is also used in electrical and electronic applications such as computer components.

Aromatic polyamides are known as *polyaramids*. There are two established fibres in this class, i.e. poly(*m*-phenylene terephthalamide) (trade name 'Nomex'):

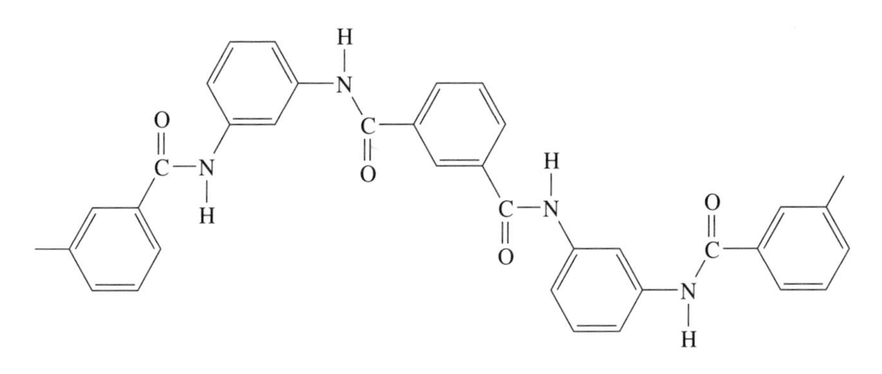

and poly(p-phenylene terephthalamide) (trade name 'Kevlar'):

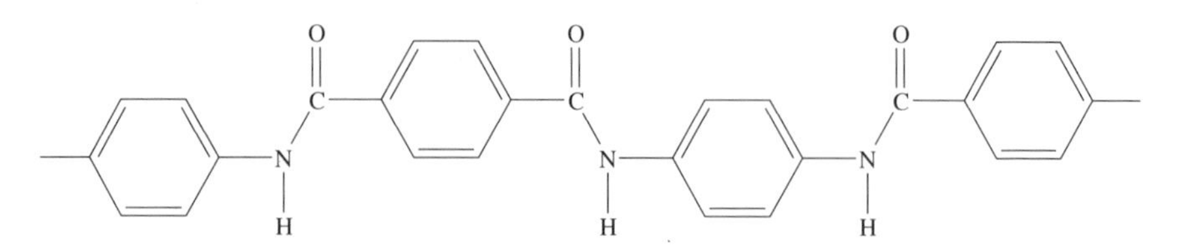

Nomex fibres were introduced on to the market because of their superior heat resistance. The introduction of Kevlar followed because this polyamide is more readily crystallized and oriented as a result of its *para*-structure, while still maintaining excellent thermal stability. In addition, Kevlar shows unusually high tensile properties. Consequently, Kevlar fibres have been used for applications including protective clothing, ropes, composites, sporting goods and aeronautical engineering components.

Poly(ether ether ketone) (PEEK) (trade name 'Victrex') is a thermoplastic polymer with the following structural repeat unit:

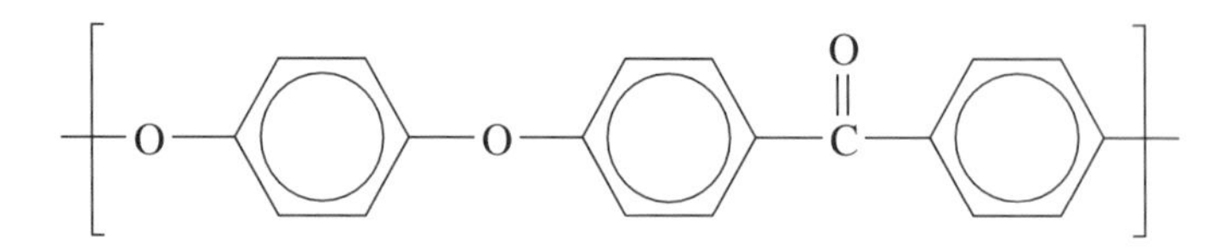

This HPP is of use in bearing-type applications because of its good wear properties. Although this polymer is comparatively expensive, PEEK has also found wide application in composites and the aerospace industry.

DQ 1.4

Suggest an HPP that would be suitable for use in a tennis racquet frame.

Answer

Kevlar fibres are employed in composite materials used to produce sporting goods, such as tennis racquets, because it is lightweight and possesses excellent mechanical properties. Polymer composites are discussed further below in Section 1.9.

1.7 Copolymers

Copolymers are comprised of chains containing two or more different types of monomers [4, 5]. Even in the simplest case of a copolymer containing two monomer groups, there are a number of possible copolymer types. The most commonly formed copolymer is a *random copolymer*, where the monomers are arranged randomly along the polymer chain. If the component monomers are denoted as A and B, then the chain would appear somewhat like –ABBAAABAABBBA–. When the monomers alternate in a regular manner in the polymer chain (–ABABABAB–), an *alternating copolymer* is formed. *Block copolymers* contain long sequences of one monomer joined to another long sequence of the other monomer (–AAAAABBBBB–). A *graft polymer* is formed by attaching chains of one monomer to the main chain of another homopolymer. The general structure of graft copolymers is as follows:

```
—AAAAAAAAA—
     B
     B
     B
     B
```

The nomenclature for copolymers is summarized in Table 1.4.

There are a number of copolymers of commercial importance and the properties and applications of some common copolymers are summarized in Table 1.5. *Styrene–acrylonitrile* (SAN) resins are random copolymers of styrene and acrylonitrile. SAN resins are tough and chemically resistant with good mechanical properties as a result of copolymerization creating hydrogen bonding between the polymer chains. SAN blends have the advantages of clarity, rigidity, hardness and easy processability. Consequently, such resins have been employed in automotive, medical and household applications where clarity is required.

The most widely used synthetic elastomer is *styrene–butadiene rubber* (SBR). This rubber is a styrene–butadiene copolymer and usually contains about 20% styrene. The copolymer can be vulcanized with sulfur because of the presence of butadiene. The butadiene units ($CH_2{=}CHCH{=}CH_2$) form a mixture of 1,2- and 1,4-, and *cis*- and *trans*-isomers. The styrene allows for a tough and strong rubber, while the styrene benzene group reduces the possibility of crystallization. SBR is used for tyres because of its better wear resistance and lower cost than natural rubber. However, SBR blends do have the disadvantage that they can absorb oils.

Table 1.4 Copolymer nomenclature

Copolymer	Nomenclature
Unspecified	Poly(A-*co*-B)
Statistical	Poly(A-*stat*-B)
Alternating	Poly(A-*alt*-B)
Random	Poly(A-*rand*-B)
Block	Poly A-*block*-poly B
Graft	poly A-*graft*-poly B

Table 1.5 The properties and applications of some common copolymers

Name	Trade names	Properties	Applications
Styrene–acrylonitrile (SAN)	Luran	Tough, good chemical resistance, good mechanical properties, rigid, easily processed	Automotive, medical and household applications
Styrene–butadiene rubber (SBR)	Polysar	Tough, good wear resistance, inexpensive, poor oil resistance	Tyres, shoe soles, flooring, hoses
Acrylonitrile–butadiene (NBR)	Nitrile, Chemigum, Krynak	Good oil and abrasion resistance	Seals, hoses, shoe soles

Nitrile rubbers are copolymers of butadiene (55–82 wt%) and acrylonitrile (45–18 wt%). The nitrile helps provide good oil, solvent and abrasion resistance. Although nitrile rubbers tend to be more expensive than other synthetic rubbers, they do find use in hoses and gaskets, where high oil and solvent resistance is required.

SAQ 1.3

A copolymer consists of 35 wt% acrylonitrile and 65 wt% styrene. Determine the mole fraction of the two components of this copolymer.

1.8 Blends

Polymer blends are defined as *mixtures of polymeric materials* and consist of at least two polymers or copolymers [6, 7]. The main reason for blending is economy. For instance, the performance of an engineering resin can be extended by diluting it with a lower-cost polymer. In this way, materials with a full set of desired properties can be developed. In addition, industrial polymer scrap may also be recycled.

Acrylonitrile–butadiene–styrene (ABS) blends currently dominate the market. ABS blends are composed of two copolymers, with a matrix consisting of a styrene–acrylonitrile copolymer and a rubbery phase consisting of a styrene–butadiene copolymer. ABS blends possess good impact and mechanical strength, with their useful mechanical properties being due to the contributing properties of each of their components. Acrylonitrile provides chemical resistance, heat resistance and toughness, while butadiene provides impact strength and styrene provides rigidity and easy processing. ABS blends are used for piping, electrical and automotive engineering. Some examples of commercially important polymer blends are PC/ABS, ABS/PVC, ABS/PET and PBT/PC.

1.9 Composites

Composites are materials composed of a mixture of two or more phases [3, 8]. Polymers are employed in a wide range of important composites. For example, glass-reinforced polyester is a polymer composite consisting of glass fibres embedded in a polyester thermosetting resin. Composites are increasingly replacing metals in many applications, in particular engineering, where the main advantage is weight reduction with improved mechanical properties.

The fibre component of polymer composites reinforces the thermosetting matrix. Fibres carry most of the load because they are chosen for their high tensile modulus and strength. The fibres also toughen the brittle matrix by blocking or deflecting any cracks that may propagate through the polymer. In addition, the fibres can be selectively arranged to lie in specific directions and locations where the maximum stresses are likely to occur. A common fibre used in composites is Kevlar.

The matrix of a polymer composite may be composed of epoxy resins or polyester resins. The role of the matrix resin is to determine and maintain the shape of the composite, to keep the fibres in position, to prevent the fibres from buckling, and to protect the fibre surfaces from chemical and mechanical damage. A schematic of a cross-section of a polymer composite is shown in Figure 1.4.

The composition of a composite may be defined in terms of the composite density (ρ_c), which is given by the following equation:

$$\rho_c = V_m \rho_m + V_f \rho_f \tag{1.1}$$

where V_m is the matrix volume fraction, ρ_f is the fibre density, V_f is the fibre volume fraction ($V_f = 1 - V_m$), and ρ_m is the matrix density.

SAQ 1.4

A sample of a glass-filled nylon 6,6 composite is found to have a density of 1515 kg m^{-3}. The densities of nylon 6,6 and glass are 1135 and 2500 kg m^{-3}, respectively. What is the volume fraction of glass fibre in the nylon ?

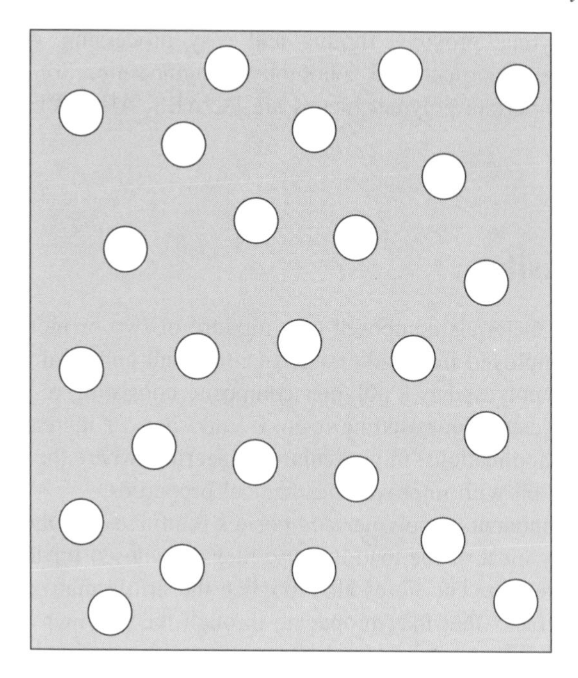

Figure 1.4 Schematic of a cross-section of a polymer composite.

1.10 Additives

Certain types of substances are introduced into polymers during their manufacture in order to improve their properties [9]. These substances are known as *additives* and may include fillers, colourants, plasticizers, stabilizers and flame retardants. *Fillers* are added to polymers in order to improve certain mechanical properties, such as strength and toughness, and usually consist of materials such as calcium carbonate, glass, clay or fibres. *Colourants*, such as dyes or pigments, are used to impart a specific colour. Polymers may have their flexibility improved by the addition of *plasticizers*. These tend to be low-molecular-weight liquids such as phthalate esters. *Stabilizers* are used to counteract the degradation of polymers by exposure to light and oxygen and include lead oxide, amines and carbon black. The flammability may also be minimized by the addition of *flame retardants* such as antimony trioxide.

1.11 Speciality Polymers

There is an abundance of everyday applications for polymers. However, there is also a wide range of specialist applications for such materials – some of these are summarized in the following.

1.11.1 Liquid Crystalline Polymers

Liquid crystalline polymers (LCPs) are so named because they display a liquid crystalline state which is neither strictly crystalline nor liquid [10]. The structures of these polymers are discussed in more detail in Chapter 5. While other polymers are randomly oriented in the liquid state, LCP molecules are capable of becoming aligned in highly ordered conformations. Typical polymers that exhibit liquid crystalline behaviour consist of long thin molecules. In order for polymers to behave as liquid crystals, they must possess some structural rigidity. Polyaramids, such as poly(*p*-phenylene terephthalamide), and thermotropic polyesters demonstrate suitable rigidity. The main application of LCPs is in liquid crystal displays (LCDs) for computers and watches.

1.11.2 Conducting Polymers

Conventional polymers are generally poor conductors of electricity and advantage is often made of their insulating properties. However, there are some polymers with highly conjugated structures that show unusual conducting properties for such materials [11]. Polymers with conjugated π-electrons backbones can be

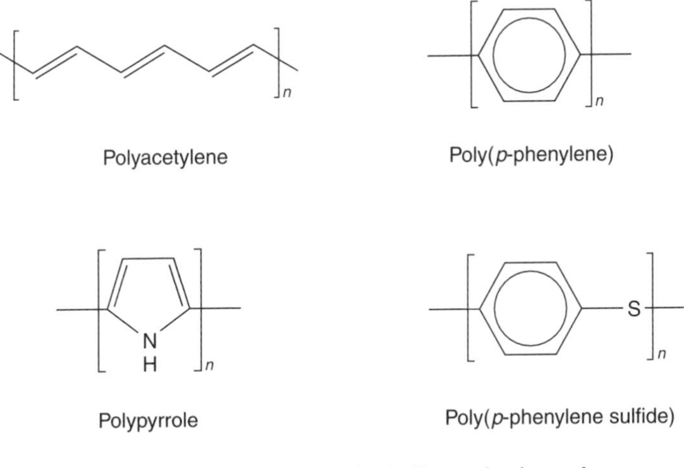

Figure 1.5 Examples of some intrinsically conducting polymers.

Table 1.6 Conductivity of intrinsically conducting polymers

Polymer	Conductivity(S cm^{-1})
Polyacetylene (PA)	$10^{-6} - 10^{4}$
Poly(*p*-phenylene) (PPP)	$10^{-15} - 5 \times 10^{2}$
Polypyrrole (PPY)	$10^{-4} - 2 \times 10^{3}$
Poly(*p*-phenylene sulfide) (PPS)	$10^{-16} - 1$

doped to produce materials that exhibit electrical conductivities approaching those of metals. Materials such as polyacetylene, poly(*p*-phenylene), polypyrrole and poly(*p*-phenylene sulfide) have been identified as *intrinsically conducting polymers* (ICPs). The structures of these polymers are shown in Figure 1.5. A number of different charge-transfer agents, known as dopants, have been used to improve the conductivity of such polymers. Substances such as AsF_5, I_2 and BF_3 have been employed as dopants. The doping of conjugated polymers generates high conductivities by increasing the carrier concentration via oxidation or reduction with electron acceptors or donors, respectively. Table 1.6 lists the conductivities of some common polymers. Note that the value of the conductivity depends on the nature and the concentration of the dopant which is present.

1.11.3 Thermoplastic Elastomers

Thermoplastic elastomers (TPEs) are polymers that, at ambient temperatures, exhibit elastomeric behaviour while being fundamentally thermoplastic in character [12]. As already discussed, most elastomers are cross-linked and do not show the melting behaviour of thermoplastics. However, elastomeric and thermoplastic behaviour may be combined in a TPE by producing a block copolymer of a hard, rigid thermoplastic and a soft, flexible elastomer. The two block types occupy alternate positions, with hard segments generally located at the chain ends, and a soft central region. The hard domains act as physical cross-links at room temperature. Examples of TPEs are styrenic block copolymers, which may contain butadiene or isoprene as the rubbery component. Figure 1.6 represents the molecular structure of a TPE. One of the main advantages of these materials is that when they are heated above the melting temperature of the hard phase,

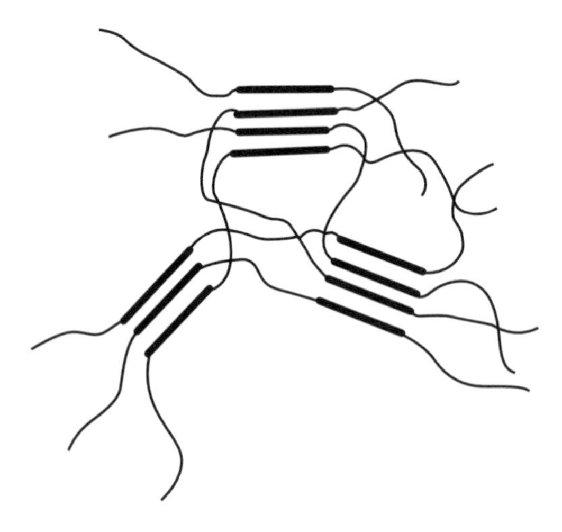

Figure 1.6 Structure of a thermoplastic elastomer.

they melt because the physical cross-links are disrupted. This allows TPEs to be easily processed.

1.11.4 Biomedical Polymers

Polymers have found widespread use in biomedical applications [13]. The lightness, relative inertness and easy processing properties of polymers make them attractive materials for such applications. However, there are some general requirements for biomedical polymers. The material must not leach or release soluble components into a living system unless *such release is intentional*, such as in drug delivery. A living system must not degrade the implant unless *degradation is intentional*, such as for an absorbable suture. The *mechanical* and *physical* properties of the polymer *must be appropriate* for the intended function. For example, a tendon replacement must have high tensile strength. Furthermore, the polymer must be *biocompatable* (e.g. not carcinogenic, does not cause inappropriate immunological response, etc.), *sterilizable* and *free of bacteria and toxins*. Conventionally manufactured polymers rarely satisfy these criteria and, in general, materials must be specifically engineered to meet stringent demands. Such polymers have found use in cardiac, opthalomological, dental and orthopaedic applications, and in drug delivery. Table 1.7 summarizes the types of polymers used for some specific biomedical applications.

Table 1.7 Biomedical applications of polymers

Application	Polymers
Cardiac	
heart valves	PUs, PVC, silicones
pacemakers	PUs, silicones
vascular grafts	PUs, PTFE, polyesters
Opthalmological	
contact lenses	PHEMA[a], silicones, acrylics
corneal implants	PHEMA
sutures	Poly(glycolic acid), nylon, PP, polyesters
Dental	
dentures	Natural rubber, PMMA, PC, nylon
facial surgery	Silicones
artificial teeth	PMMA
adhesives	Poly(acrylic acid)
Orthopaedic	PMMA, UHMWPE, silicones
Plastic surgery	Silicones
Drug delivery	Poly(lactic acid), polycaprolactone, polyorthoesters, polyacrylamide, PVAl

[a] PHEMA, poly(hydroxyethyl methacrylate).

1.11.5 Biodegradable Polymers

Biodegradable polymers are materials that can be degraded by microorganisms and enzymes. The use of such polymers provides an approach to the problem of plastic waste. Biodegradable polymers can also be used for medical applications such as implants, sutures and in drug release, and for agricultural applications such as mulch and agrochemicals [14–18]. Polymers that are biologically degraded contain functional groups that are susceptible to enzymatic hydrolysis and oxidation. Polyesters, PVAl, PUs and poly(vinyl ethanoate) are examples of such materials. Biodegradable polymers need to be designed to have a certain lifetime and then have degradation triggered by, for instance, exposure to ultraviolet radiation.

Summary

In this chapter, the fundamental structures displayed by polymers were introduced. A brief history of the development of synthetic polymers was then outlined. The structures, properties and applications of common thermoplastics, thermosets, elastomers and high-performance polymers were described. Copolymers, blends and composites were also defined and discussed, and the common additives used in polymers were outlined. In addition, some speciality applications of polymers, including liquid crystalline polymers, conducting polymers, thermoplastic elastomers, biomedical polymers and biodegradable polymers, were described.

References

1. Fenichell, S., *Plastic: The Making of a Synthetic Century*, Harper Collins, New York, 1996.
2. Billmeyer, F. W., *Textbook of Polymer Science*, 3rd Edn, Wiley, New York, 1984.
3. Baer, E. and Moet, A., *High Performance Polymers*, Carl Hanser Verlag, Munich, 1991.
4. Hamielec, A. E., MacGregor, J. E. and Penlidis, A., 'Copolymerization', in *Comprehensive Polymer Science*, Vol. 3, Ledwith, A., Russo, S. and Sigwatt, P. (Eds), Pergamon Press, Oxford, UK, 1989, pp 17–32.
5. Cowie, J. M. G., 'Block and Graft Copolymerization', in *Comprehensive Polymer Science*, Vol. 3, Ledwith, A., Russo, S. and Sigwatt, P. (Eds), Pergamon Press, Oxford, UK, 1989, pp. 33–42.
6. MacKnight, W. J. and Karasz, F. E., 'Polymer Blends', in *Comprehensive Polymer Science*, Vol. 7, Aggarwal, S. L. and Russo, S. (Eds), Pergamon Press, Oxford, UK, 1989, pp. 111–130.
7. Walsh, D. J., 'Polymer Blends', in *Comprehensive Polymer Science*, Vol. 2, Booth, C. (Eds), and Price, C. Pergamon Press, Oxford, UK, 1989, pp. 135–154.
8. Keeny, J. M. and Nicolais, L., 'Science and Technology of Polymer Composites', in *Comprehensive Polymer Science*, 1st Suppl., Aggarwal, S. L. and Russo, S. (Eds), Pergamon Press, Oxford, UK, 1992, pp. 471–526.
9. Gachter, R. and Muller, H., *Plastics Additives Handbook*, Hanser Publishers, Munich, 1987.
10. Percec, V. and Tomazos, D., 'Molecular Engineering of Liquid Crystalline Polymers', in *Comprehensive Polymer Science*, 1st Suppl., Aggarwal, S. L. and Russo, S. (Eds), Pergamon Press, Oxford, UK, 1992, pp. 299–384.

11. Chandrasekhar, P., *Conducting Polymers, Fundamentals and Applications: A Practical Approach*, Kluwer Academic, Boston, 1999.
12. Dyson, R. W., *Engineering Polymers*, Blackie, Glasgow, UK, 1990.
13. Ratner, B. D.. 'Biomedical Applications of Synthetic Polymers', in *Comprehensive Polymer Science*, Vol. 7, Aggarwal, S. L. and Russo, S. (Eds), Pergamon Press, Oxford, UK, 1989, pp. 201–248.
14. Scott, G.. *Polymers and the Environment*, Royal Society of Chemistry, Cambridge, UK, 1999.
15. Albertsson, A. C. and Karlsson, S.. 'Biodegradable Polymers', in *Comprehensive Polymer Science*, 1st Suppl., Aggarwal, S. L. and Russo, S. (Eds), Pergamon Press, Oxford, UK, 1992, pp. 285–298.
16. Huang, S. J., 'Biodegradation', in *Comprehensive Polymer Science*, Vol. 6, Eastmond, G. C., Ledwith, A., Russo, S. and Sigwalt, P. (Eds), Pergamon Press, Oxford, UK, 1989, pp. 597–606.
17. Williams, D. F., 'Polymer Degradation in Biological Environments', in *Comprehensive Polymer Science*, Vol. 6, Eastmond, G. C., Ledwith, A., Russo, S. and Sigwalt, P. (Eds), Pergamon Press, Oxford, UK, 1989, pp. 607–620.
18. Gandini, A., 'Polymers from Renewable Sources', in *Comprehensive Polymer Science*, 1st Suppl., Aggarwal, S. L. and Russo, S. (Eds), Pergamon Press, Oxford, UK, 1992, pp. 527–574.

Chapter 2

Identification

Learning Objectives

- To carry out and interpret basic identification methods, including solubility, density and heating tests, on polymers.
- To use infrared spectroscopy, Raman spectroscopy, nuclear magnetic resonance spectroscopy, ultraviolet–visible spectroscopy, differential scanning calorimetry, mass spectrometry, chromatographic techniques and emission spectroscopy to identify polymers.

2.1 Introduction

There are various standard techniques used to identify the chemical structures of polymers. As a starting point, the common methods for functional groups and elemental analysis used in analytical chemistry are applicable. Spectroscopic techniques, including infrared, Raman, ultraviolet–visible, nuclear magnetic resonance and emission spectroscopies, are regularly used by polymer scientists. These techniques are introduced in this chapter. In addition, thermal analysis, e.g. differential scanning calorimetry (DSC), is an important identification technique for polymers and is also discussed in the following. Mass spectrometry and chromatographic techniques, including gas chromatography, high performance liquid chromatography and thin layer chromatography, can also be used to separate components and may be used to identify both monomers and additives.

2.2 Preliminary Identification Methods

There are a number of simple methods that can be used to check types of polymers. These include solubility tests, melting temperature determination, density measurements, flame tests and pyrolysis tests. There are also numerous standard chemical tests that can be used to identify the functional groups present in various polymers. These will not be detailed here, but are well described elsewhere [1, 2].

2.2.1 *Solubility*

A check of polymer solubility provides a simple method of distinguishing polymers in the first instance. Solubility can be determined by placing approximately 0.1 g of the polymer of interest in a test-tube with about 5 ml of a particular solvent. As some polymers may take a longer period of time to dissolve (or swell), it is advisable to check the sample after standing for several hours. The tube may also require heating to induce solubility. Table 2.1 summarizes the solvents and non-solvents for a number of common polymers at room temperature.

Table 2.1 Solvents and non-solvents for various common polymers

Polymer	Solvent	Non-solvent
PE	Hydrocarbons[a]	All liquids at room temperature
PP	Hydrocarbons	Ethyl acetate
PVC	Dimethylformamide, tetrahydrofuran, cyclohexanone, methyl ethyl ketone	Hydrocarbons, acetone
PS	Cyclohexane, benzene, methyl ethyl ketone, toluene	Aliphatic hydrocarbons, alcohols, acetone
PMMA	Benzene, chloroform, acetone, aromatic hydrocarbons	Aliphatic hydrocarbons, alcohols
Nylon(s)	*m*-Cresol, acetone	Hydrocarbons, ether
PTFE	None	All
PAN	Dimethylformamide	Hydrocarbons, alcohols, ketones, ether
Cellulose acetate	Chloroform, acetone	Hydrocarbons, acetone
PVA	Methanol, benzene, chloroform, acetone	Aliphatic hydrocarbons
PVAl	Water, dimethyl formamide	Hydrocarbons, ketones, methanol
PET	Phenol	Hydrocarbons, acetone
PU	Formic acid, *m*-cresol	Methanol, ether, hydrocarbons
Natural rubber	Hydrocarbons, chlorinated hydrocarbons, ether	Alcohols, acetone
PDMS	Chloroform, benzene, ether	Methanol, ethanol
Polychloroprene	Benzene, chlorobenzene	Aliphatic hydrocarbons, alcohols, ketones
SBR	Benzene, methyl ethyl ketone	Alcohols
Nitrile rubber	Benzene	Methanol

[a] At temperatures $> 80°C$.

2.2.2 Density

The density of a material (ρ) is the ratio of the mass (m) to the volume (V), as follows:

$$\rho = m/V \tag{2.1}$$

The simplest method for determining the density of a polymer is the use of a flotation procedure. This approach involves floating the polymer sample in a liquid of known density. For instance, samples can be immersed in columns of methanol ($\rho = 0.79$ g cm^{-3}), water ($\rho = 1.00$ g cm^{-3}), saturated aqueous magnesium chloride solution ($\rho = 1.34$ g cm^{-3}) or saturated aqueous zinc chloride solution ($\rho = 2.01$ g cm^{-3}). The behaviour of the samples is then observed, i.e. whether the sample remains on the liquid surface, floats within the liquid or sinks. The specific behaviour determines whether the polymer has a lower or higher density than the liquid in which it is immersed. Table 2.2 lists the densities of some common polymers.

2.2.3 Behaviour on Heating

Heating a polymer sample can provide a number of preliminary approaches to analysis. *Flame tests* can be used to differentiate polymers, as the flame produced by a burning polymer shows characteristic differences dependent on the structure of the material. In order to test the behaviour of a polymer in a flame, a small piece of the material can be held with a spatula or a pair of tweezers in the low flame of a Bunsen burner. The appearance of the flame and the production of any odours should be noted. Table 2.3 summarizes the results of flame tests for a number of common polymers.

The degradation behaviour of polymers can be studied by using *pyrolysis tests*. Thermal degradation produces low-molecular-weight fragments that are often flammable or have a characteristic odour. A polymer sample can be heated without direct contact with the flame by placing a small piece of the material in a pyrolysis tube and holding the tube with tongs in a Bunsen flame. At the open end of the tube, a piece of moist litmus or pH paper is held in order to determine the nature of the fumes produced. The polymer can be distinguished as acidic, basic or neutral. Table 2.4 shows groupings of some common polymers into these categories. Although many polymers are fairly neutral, certain types are basic or acidic in nature.

For thermoplastics, the *melting temperature* may be used to identify the polymer. The melting behaviour can be observed by using a standard melting point tube or with a hot-stage. The latter device provides a temperature gradient which is created by resistance heaters along a metal bar. Small pieces of sample are placed directly at different points on the metal bar. The temperature at the border between solid and molten material is determined directly from the scale on the hot-stage. The melting temperatures (or ranges) for some common polymers are listed in Table 2.5.

Table 2.2 Densities of various common polymers

Polymer	Density (g cm^{-3})
Silicone rubber	0.80
PP	0.85–0.92
LDPE	0.89–0.93
Natural rubber	0.92–1.00
UHMWPE	0.94
HDPE	0.94–0.98
Nylon 12	1.01–1.04
Nylon 11	1.03–1.05
PS	1.04–1.06
ABS	1.04–1.08
SAN	1.06–1.10
Nylon 6,10	1.07–1.09
Polyester resins	1.10–1.40
Epoxy resins	1.10–1.40
Nylon 6	1.12–1.15
Nylon 6,6	1.13–1.16
PAN	1.14–1.17
PVA	1.17–1.20
Nylon 4,6	1.18
PMMA	1.16–1.20
PC	1.20–1.22
PU	1.20–1.26
PVAl	1.21–1.31
Cellulose acetate	1.25–1.35
PEEK	1.26–1.32
PEI	1.27
P–F resins	1.26–1.28
PET	1.38–1.41
PBT	1.31
Cellulose nitrate	1.34–1.40
PES	1.37
Unplasticized PVC	1.38–1.41
Polyimide	1.42
Kevlar	1.44
Amino resins	1.47–1.52
PPS	1.66
PVDF	1.76
PTFE	2.10–2.30

Table 2.3 Flame test results for some common polymers

Polymer	Colour	Odour
PE	Yellow, blue base	Burning candle
PP	Yellow, blue base	Burning candle
PVC	Yellow, green base	Acrid
PS	Yellow, blue base, sooty	Styrenic
PMMA	Yellow, blue base	Sweet
Nylon(s)	Yellow, blue smoke	None
PTFE	Yellow	None
PAN	Yellow	Burning wood
Cellulose acetate	Yellow	Acetic acid
Cellulose nitrate	Yellow	Camphor
PVA	Yellow	Vinyl acetate
PVAl	Shiny	None
PET	Yellow, sooty	Sweet
Epoxy resin	Orange/yellow, sooty	Acrid
Amino resins	Yellow/blue, green edge	Formaldehyde
Polyester resin	Yellow, blue base, sooty	Styrenic
PU	Yellow, blue base	Acrid
Natural rubber	Yellow, sooty	Pungent
Silicone rubber	Bright yellow, white	None
Polychloroprene	Yellow, sooty	Acrid
PC	Yellow, sooty	Phenol
ABS	Yellow, sooty	Styrenic

Table 2.4 Pyrolysis test categories for some common polymers

pH 0.5–4.0	pH 5.0–5.5	pH 8.0–9.5
Polyester resin	Silicones	Nylon(s)
PU elastomers	PP	PAN
Cellulose acetate	PE	Amino resins
Cellulose nitrate	PS	Phenolic resin
PET	Epoxy resin	ABS
PVC	PVA	
PTFE	PVAl	
	PMMA	
	PC	
	Cross-linked PU	

Table 2.5 Melting temperatures (or ranges) for some common polymers

Polymer	Temperature (or range) (°C)
PDMS	−54
Natural rubber	28
PVA	35–85
Chloroprene	80
LDPE	115
PVDF	115–140
UHMWPE	120
PMMA	120–160
PP	160–170
LLDPE	125
Cellulose acetate	125–175
HDPE	130–137
Nylon 12	170–180
Unplasticized PVC	175
Nylon 11	180–190
Nylon 6,10	210–220
Nylon 6	215–225
PBT	220–267
PS	240
PET	245–265
Nylon 6,6	250–260
PC	265
PVAl	265
PPS	285
PAN	317
PTFE	327
PEEK	334
Polyaramid	640

SAQ 2.1

Some preliminary tests were carried out on a sample of a commercial polymer of unknown content. Use the test results listed in Table 2.6 below to identify the major polymer component of the sample.

Table 2.6 Preliminary identification test results for a sample of a commercial polymer of unknown content (SAQ 2.1)

Test	Result
Solubility	Soluble in chloroform and acetone, insoluble in methanol
Density	1.19 g cm^{-3}
Flame test	Yellow flame with blue base, sweet odour
Pyrolysis	Neutral pH
Melting temperature	∼130°C

2.3 Infrared Spectroscopy

Infrared (IR) spectroscopy is a popular method for characterizing polymers [3,4]. This technique is based on the vibrations of the atoms of a molecule. An infrared spectrum is obtained by passing infrared radiation through a sample and determining which fraction of the incident radiation is absorbed at a particular energy. The energy at which any peak in an absorption spectrum appears corresponds to the frequency of vibration of a part of the sample molecule.

Most infrared spectroscopy is carried out by using *Fourier-transform infrared* (FTIR) spectrometers. This method is based on the interference of radiation between two beams to yield an *interferogram*, i.e. a signal produced as a function of the change of pathlength between the two beams. The two domains of distance and frequency are interconvertible by the mathematical method of *Fourier transformation*. The basic components of an FTIR spectrometer are shown schematically in Figure 2.1. The radiation emerging from the source is passed through an interferometer to the sample before reaching a detector. Upon amplification of the signal, in which high-frequency contributions have been eliminated by a filter, the data are converted to a digital form by using an analog-to-digital converter and then transferred to the computer for Fourier transformation to take place.

There are a number of methods available for examining polymer samples. If the polymer is a thermoplastic it can be softened by warming and pressed into a thin film by using an hydraulic press. Alternatively, the polymer can be dissolved in a volatile solvent and the solution allowed to evaporate to a thin film on an alkali halide plate. Some polymers, such as cross-linked synthetic rubbers, can be microtomed, i.e. cut into thin slices with a blade. A solution in a suitable solvent is also a possibility. If the polymer is a surface coating, reflectance techniques may be used. Reflectance sampling methods are of particular assistance when studying polymer surfaces and are discussed in further detail in Chapter 6.

The output from an infrared instrument is referred to as a *spectrum*. Inverse wavelength units (cm^{-1}) are used on the x-axis – this is known as the *wavenumber* scale. The y-axis may be represented by *% transmittance*, with 100% at the top of the spectrum. It is commonplace to have the choice of *absorbance* or *transmittance* as a measure of band intensity. The transmittance is traditionally used for spectral interpretation, while absorbance is used for quantitative work.

Figure 2.1 Schematic of a typical Fourier-transform infrared (FTIR) spectrometer.

The infrared spectrum can be divided into three regions, namely the *far-infrared* (< 400 cm^{-1}), the *mid-infrared* (400–4000 cm^{-1}) and the *near-infrared* (4000–13 000 cm^{-1}). Most infrared applications employ the mid-infrared region, although the near- and far-infrared regions can also provide specific information about materials. The near-infrared region consists largely of overtones or combination bands of fundamental modes appearing in the mid-infrared region. The far-infrared region can provide information regarding lattice vibrations.

Spectrum interpretation is simplified by the fact that the bands that appear can be assigned to particular parts of the molecule, thus producing what are known as *group frequencies*. The mid-infrared spectrum may be divided into the following four regions:

- X–H stretching region (4000–2500 cm^{-1})
- triple-bond region (2500–2000 cm^{-1})
- double-bond region (2000–1500 cm^{-1})
- fingerprint region (1500–600 cm^{-1})

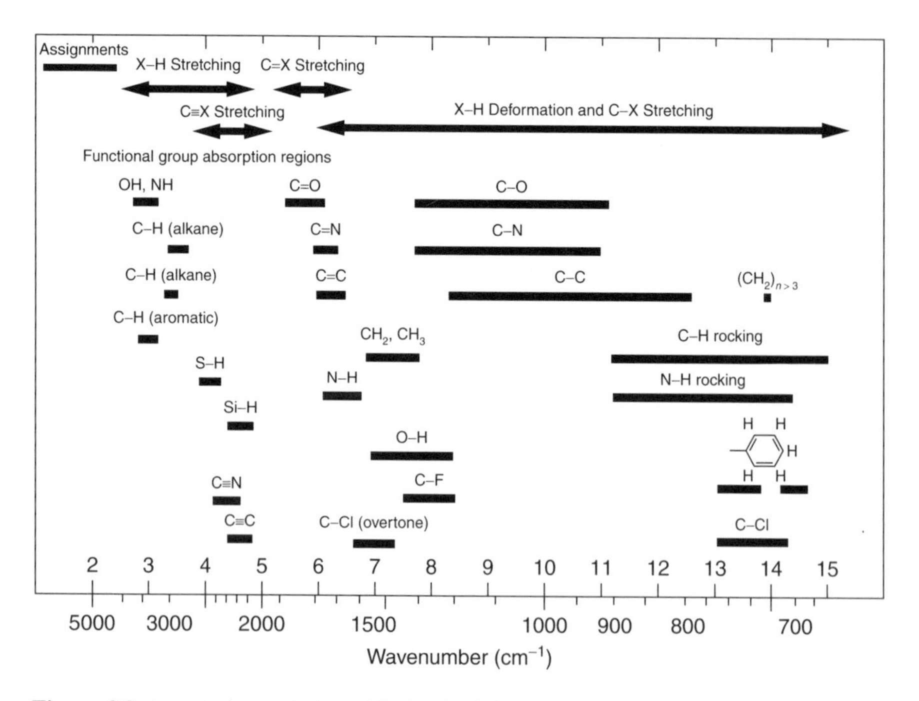

Figure 2.2 A typical correlation table for the infrared modes of polymers. From Sandler, S. R., Karo, W., Bonesteel, J. and Pearce, E. M., *Polymer Synthesis and Characterization: A Laboratory Manual*, © Academic Press, 1998. Reproduced by permission of Academic Press.

The information provided by these regions can be summarized in what are known as *correlation tables*. A useful correlation table for polymers is shown in Figure 2.2. The information obtained from these tables can be combined with libraries of infrared spectral data [5].

SAQ 2.2

Polycaprolactone (PCL) has the following structural repeat unit:

$$-\!\!\left[O - (CH_2)_5 - CO \right]\!\!-_n$$

The infrared spectrum of PCL is shown below in Figure 2.3. Identify and tabulate the major infrared modes for this polymer.

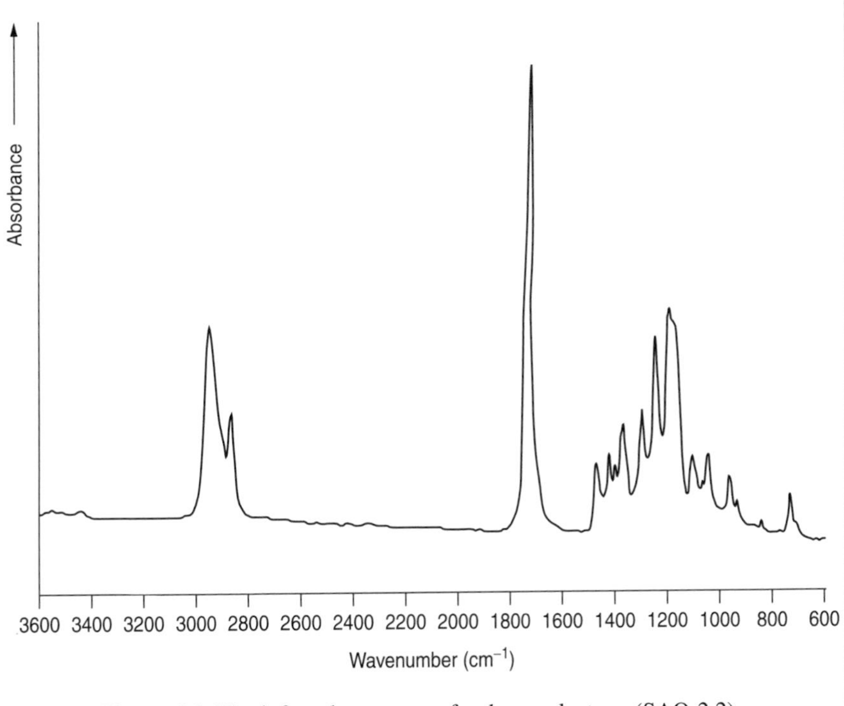

Figure 2.3 The infrared spectrum of polycaprolactone (SAQ 2.2).

SAQ 2.3

Figure 2.4 below shows an infrared spectrum of a polymer. What is the likely structure of this material?

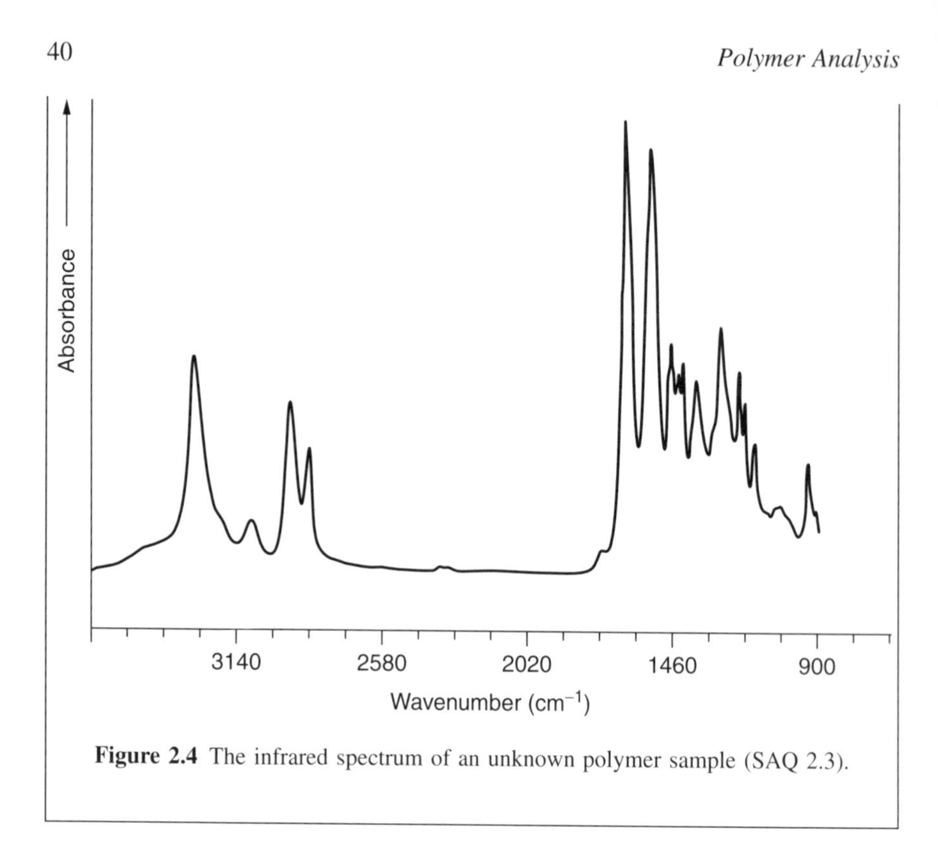

Figure 2.4 The infrared spectrum of an unknown polymer sample (SAQ 2.3).

The composition of copolymers and blends can be quantitatively determined by using infrared spectroscopy. The Beer–Lambert law (see Section 2.6 below) can be applied. Ideally, distinctive representative modes for the polymers should be identified. For example, in the case of vinyl chloride/vinyl acetate copolymers, the ratio of the absorbance of the acetate mode at 1740 cm^{-1} to that of the vinyl chloride methylene bending mode at 1430 cm^{-1} can be used for quantitative analysis. Copolymers or blends of known composition can be used for calibration purposes.

SAQ 2.4

In the infrared spectra of styrene/acrylonitrile copolymers, the ratio of the absorbances of the C≡N nitrile stretching vibration at 2250 cm^{-1} and the polystyrene (PS) ring-stretching vibration at 1600 cm^{-1} may be employed as a measure of the composition. Infrared analysis of a number of styrene/acrylonitrile copolymers of known compositions yielded the results listed in Table 2.7 below. The infrared spectrum of a sample of poly(styrene-*co*-acrylonitrile) of unknown composition was recorded. The absorbance values at 2250 and 1600 cm^{-1} in this spectrum were 0.205 and 0.121, respectively. Estimate the composition of this sample.

Table 2.7 Infrared analysis data obtained for a number of styrene/acrylonitrile copolymers (SAQ 2.4)

Acrylonitrile concentration (%)	Absorbance at 2250 cm^{-1}	Absorbance at 1600 cm^{-1}
10	0.230	0.383
20	0.223	0.177
30	0.230	0.120
40	0.235	0.0909
50	0.227	0.0701
60	0.231	0.0592

2.4 Raman Spectroscopy

Although the use of Raman spectroscopy for studies of polymers has expanded in recent years, the technique is still less popular than infrared spectroscopy, largely due to economical reasons. In addition, fluorescence of polymer samples has been a drawback in Raman spectroscopy. However, these problems have now been solved with the emergence of Fourier-transform (FT) Raman spectroscopy, a technique which utilizes an increased laser wavelength and avoids fluorescence [6–9].

Raman and infrared spectroscopies are based on the same physical process, i.e. the vibrations of the atoms of a molecule. However, the interaction between the stimulating electromagnetic radiation and the samples differs in these methods. Another way in which electromagnetic radiation may interact with a molecule is by being scattered. The scattering with a change of frequency is called *Raman scattering* and the change of frequency is equal to the frequency of one of the normal modes of vibration of the molecule.

Conventional laser Raman instruments consist of a laser source illuminating the sample, followed by a series of collection optics which gather the scattered light. The light is then passed on to a monochromator for spectral analysis. A schematic diagram of a laser Raman spectrometer is shown in Figure 2.5 [3]. Argon or krypton lasers are employed in this technique. It was not until the late 1980s that FT Raman became feasible practically, as a result of the advances in FT instrumentation [6]. The source is a Nd–YAG laser operating at 1.064 μm, well below the threshold for any common fluorescence process. The scattered light from the sample passes through a series of focusing and collection optics, and the light is then passed into a FT interferometer for modulation. A series of optical filters are then used for rejection of Rayleigh scattered light, with the light being sent to the detector.

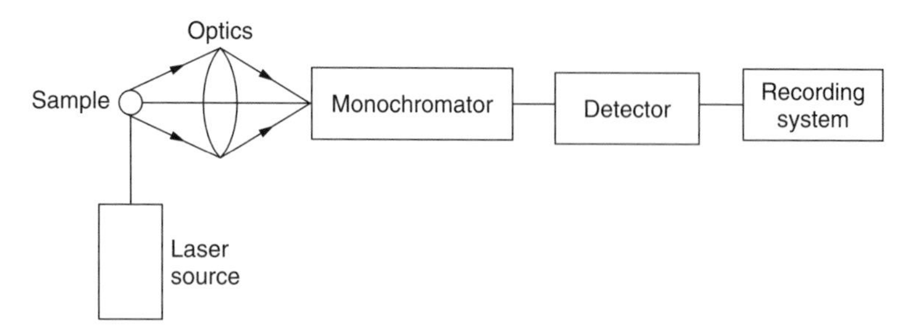

Figure 2.5 Schematic of a typical laser Raman spectrometer.

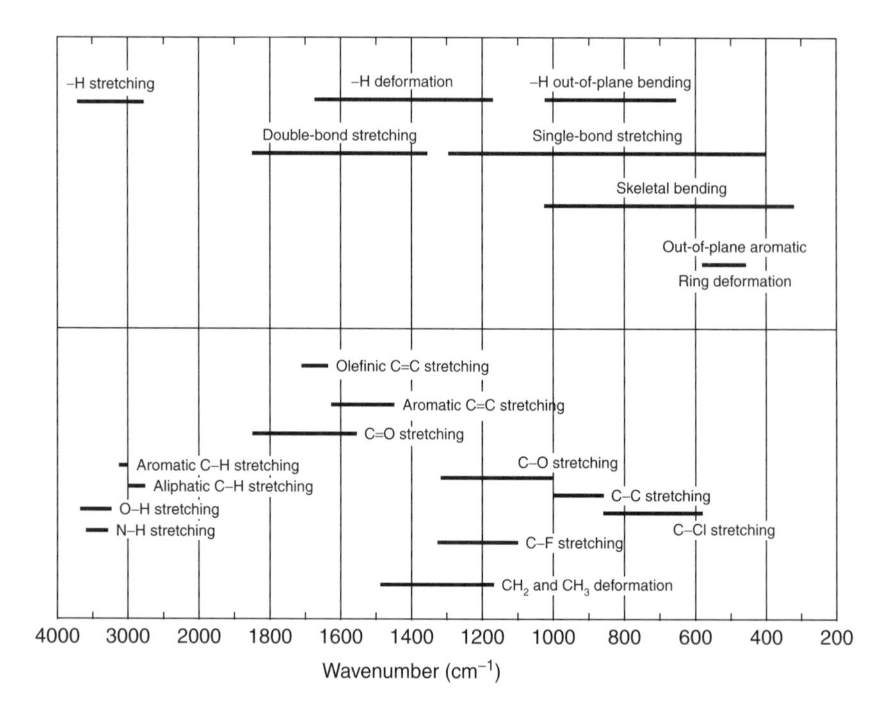

Figure 2.6 A typical correlation table for the Raman modes of polymers. From Bower, D. I. and Maddams, W. F., *The Vibrational Spectroscopy of Polymers*, © Cambridge University Press, 1989. Reproduced by permission of Cambridge University Press.

Raman spectroscopy has the advantage that a variety of sample types can be examined with minimal preparation. Solid polymers in the form of powders, films or fibres can be studied, and there is a range of appropriate solid cells commercially available for this purpose. Liquids can be examined by using silvered glass cells, and thus polymer solutions and melts can be readily investigated.

As with infrared spectroscopy, correlation tables can be used to assign Raman modes to particular functional groups. Figure 2.6 shows a correlation table of the major functional groups observed in the Raman spectra of polymers. Libraries of the Raman spectra of polymers are also available [10].

SAQ 2.5

The FT Raman spectrum of Kevlar is shown below in Figure 2.7. Identify the major Raman modes of this polymer.

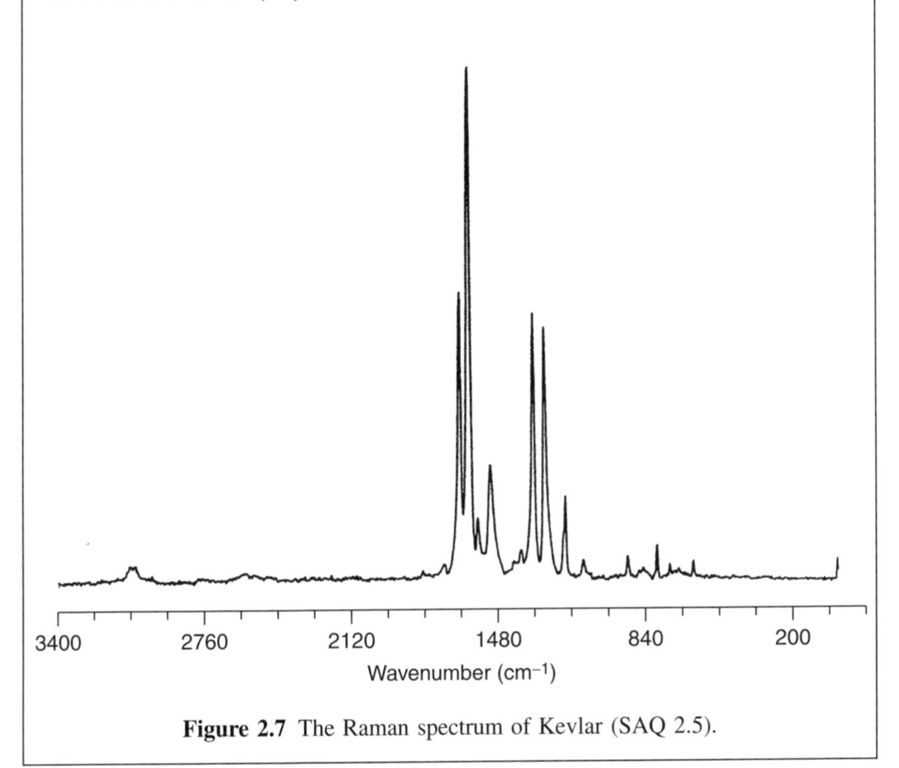

Figure 2.7 The Raman spectrum of Kevlar (SAQ 2.5).

Raman spectroscopy can be employed for the identification of nylons [11]. For instance, the Raman spectra of single-number nylons in the series from nylon 4 to nylon 12 can be used for analytical purposes. The differences in the 1700–500 cm^{-1} range, particularly for those nylons containing short methylene sequences, may be used on an empirical basis for identification. A more quantitative approach is to measure the intensity ratio of the peaks at 1440 cm^{-1} due to CH_2 bending and the amide I band (mostly C=O stretching) at 1640 cm^{-1}. This ratio increases linearly with the increasing nylon number, with a plot of the ratio as a function of the number of methylene groups being shown in Figure 2.8.

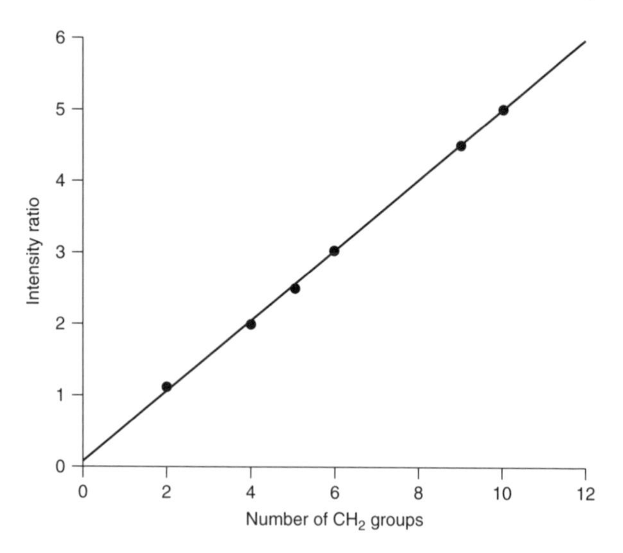

Figure 2.8 The intensity ratio of the 1440 cm^{-1}/1640 cm^{-1} modes for various *nylons* as a function of the number of methylene groups.

SAQ 2.6

The FT *Raman* spectrum of a single-number nylon is shown below in Figure 2.9. Identify the nylon type. For simplicity, use the peak heights rather than the peak areas to estimate the relevant intensity ratio.

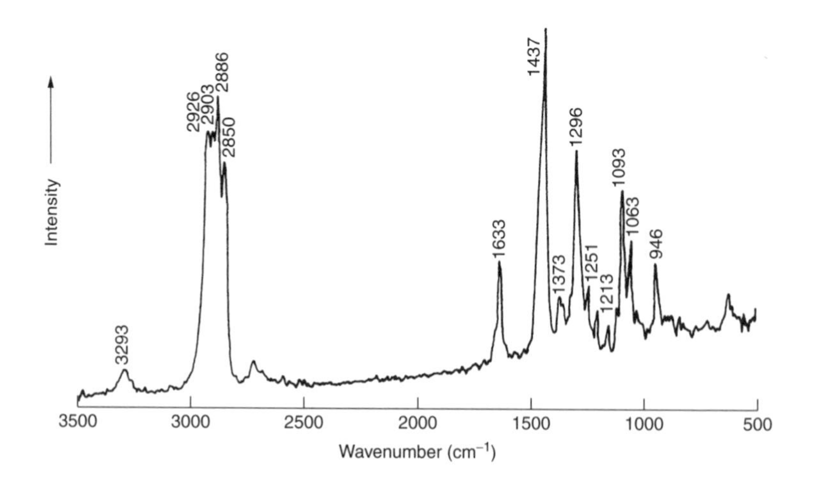

Figure 2.9 The FT Raman spectrum for a single-number nylon (SAQ 2.6). Reprinted from *Spectrochimica Acta*, **46A**, Hendra, P. J., Maddams, W. F., Royaud, I. A. M., Willis, H. A. and Zichy, V. The application of Fourier transform Raman spectroscopy to the identification and characterization of polyamides-I. Single number nylons 747–753, Copyright (1990), with permission from Elsevier Science.

Raman spectroscopy can be used to determine the composition of copolymers. Poly(ether ether sulfone) (PEES) and poly(ether sulfone) (PES) are high-performance engineering thermoplastics which may be employed in copolymer formulations. FT Raman spectroscopy has been used to characterize such copolymers [7]. The relative intensities of the phenyl ring vibrational modes of PEES and PES at 1581 and 1599 cm^{-1} provide a useful analytical method for the determination of the copolymer composition. Figure 2.10 illustrates the intensity ratio of the 1581 and 1599 cm^{-1} modes as a function of PES content in the copolymer. Two additional peaks at 1200 and 1071 cm^{-1} change in relative intensity as a function of composition, and the ratio of these bands also varies linearly. Thus, the measurement of either of these intensity ratios provides a useful analytical approach. This copolymer is amorphous so there is no problem associated with spectral changes as a result of crystallinity changes with copolymer composition.

FT Raman spectroscopy is particularly valuable for the examination of glass-filled composites as there is little interference in the Raman spectrum from the glass. The spectra of a 'pure' nylon and a nylon–glass-reinforced composite are very similar and it is possible to characterize nylon from the spectrum of the composite [7, 8]. FT Raman studies of composites containing carbon fibres have been less successful. Raman spectroscopy has limited application to *black* samples as the strong absorption of the laser radiation in the samples results in sample overheating.

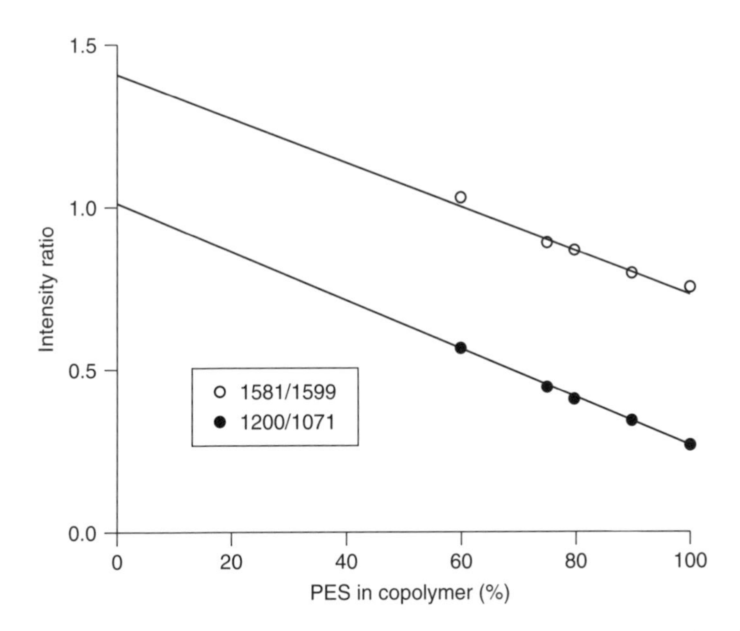

Figure 2.10 Raman intensity ratio changes of PEES/PES copolymers as a function of composition: ○, 1581 cm^{-1}/1599 cm^{-1}; ● 1200 cm^{-1}/1071 cm^{-1}.

2.5 Nuclear Magnetic Resonance Spectroscopy

Nuclear magnetic resonance (NMR) spectroscopy is a very effective tool for the study of polymer structure, both in solution and in the solid state [12–14]. NMR spectroscopy involves placing a sample in a strong magnetic field and irradiating with radiofrequency radiation. The absorptions due to the transitions between quantized energy states of nuclei that have been oriented by the magnetic field are observed.

All nuclei carry a charge and in certain nuclei this charge spins on the nuclear axis, thus generating a magnetic dipole. The angular momentum of the spinning charge may be described by using *spin numbers* (I). Each proton and neutron has its own spin, with I being the resultant of these spins. NMR spectroscopy utilizes nuclei with an $I = 1/2$, such as ^{1}H, ^{19}F, ^{13}C and ^{31}P. For such nuclei, the magnetic moments align either parallel to or against the magnetic field. The energy difference between the two states is dependent upon the *magnetogyric ratio* (γ) of the nucleus and the strength (B) of the external magnetic field. The *resonance frequency* (ν), i.e. the frequency of radiation required to effect a transition between the energy states, is given by the following expression:

$$\nu = \gamma B/2\pi \tag{2.2}$$

In theory, a single peak should be observed as a result of the interaction of radiofrequency radiation and a magnetic field on a nucleus. However, in practice absorptions at different positions are seen. This occurs because the nucleus is shielded to a small extent by its electron cloud, the density of which varies with the environment. For instance, the degree of shielding of a proton and a carbon atom observed using ^{1}H NMR spectroscopy will depend on the inductive effect of the other groups attached to the carbon atom. The difference in the absorption position from that of a reference proton is known as the *chemical shift*. A common reference compound is tetramethylsilane (TMS). The chemical shift (δ) is usually expressed in dimensionless units of ppm, determined as follows:

$$\delta = \frac{\text{frequency of absorption} \times 10^{6}}{\text{applied frequency}} \tag{2.3}$$

TMS gives single ^{1}H and ^{13}C absorptions corresponding to $\delta = 0$. Figure 2.11 illustrates the appearance of a typical NMR spectrum [15]. Each absorption area in a NMR spectrum is proportional to the number of nuclei it represents and these areas are evaluated by an integrator. The ratios of the integrations for the difference absorptions are equal to the ratios of the number of the respective nuclei present in the nucleus.

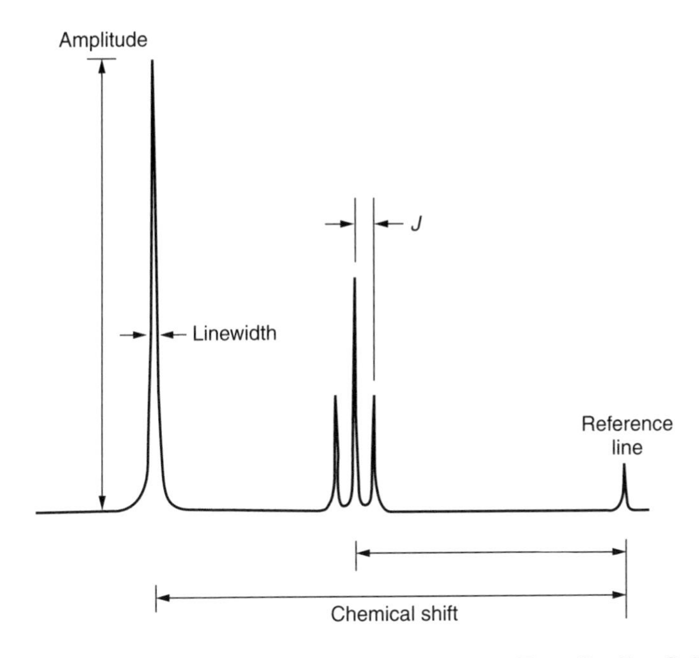

Figure 2.11 Important features of a typical NMR spectrum. From Sandler, S. R., Karo, W., Bonesteel, J. and Pearce, E. M., *Polymer Synthesis and Characterization. A Laboratory Manual*, © Academic Press, 1998. Reproduced by permission of Academic Press.

There is another phenomenon, known as *spin–spin coupling*, which further complicates NMR spectra. Spin–spin coupling is the indirect coupling of nuclei spins through the intervening bonding electrons. This occurs because there is a tendency for a bonding electron to pair its spin with the spin of the nearest $I = 1/2$ nuclei. The spin of the bonding electron so influenced will affect the spin of the other bonding electron and so on through the next $I = 1/2$ nucleus. Where spin–spin coupling occurs, each nucleus will give rise to widely separated absorptions appearing in the form of doublets as the spin of each nucleus is affected slightly by the orientations of the other nucleus through the intervening electrons. The frequency differences between the doublet peak is proportional to the effectiveness of the coupling and is known as a *coupling constant* (J) (see Figure 2.11).

While spin–spin coupling is fairly easy to interpret in proton (^{1}H) NMR, in ^{13}C NMR the spin–spin coupling between ^{1}H and ^{13}C nuclei is significant. This makes the interpretation of ^{13}C spectra difficult as in many cases the coupling is greater than the chemical shift differences between the ^{13}C nuclei absorptions. Hence, it is usual to employ *decoupling* techniques while recording ^{13}C NMR spectra. Such techniques decouple the ^{13}C spins from the ^{1}H spins so that the different ^{13}C nuclei give rise to only a single absorption in the spectrum. In addition, *two-dimensional* NMR spectra may be recorded, which can represent the different ^{13}C absorptions on one axis, while the splitting by $^{1}H-^{13}C$ spin–spin coupling is shown on the second axis (perpendicular to the first).

After absorption of energy by the nuclei during an NMR experiment, there must be a mechanism by which the nuclei can dissipate energy and return to the lower-energy state. There are two main processes, namely spin–lattice relaxation or spin–spin relaxation. *Spin–lattice relaxation* involves the transfer of energy from the nuclei to the molecular lattice, while *spin–spin relaxation* occurs from direct interactions between the spins of different nuclei that can cause relaxation without any energy transfer to the lattice.

Figure 2.12 presents a schematic diagram of the layout of an NMR spectrometer. Such spectrometers consist of a strong magnet with a homogeneous field, a radiofrequency transmitter, a radiofrequency receiver and a recorder. The instrument also contains a sample holder which positions the sample relative to the magnetic field, transmitter coil and receiver coil. The sample holder also spins the sample to increase the apparent homogeneity of the magnetic field. Polymer samples can be studied in solution and solid form. Solutions for ^{1}H and ^{13}C NMR may be prepared by using deuterated solvents such as $CDCl_3$ or C_6D_6.

Early NMR spectrometers used permanent magnets or electromagnets (e.g. 60 or 100 MHz) and are known as continuous-wave spectrometers, i.e. those in which a single radiofrequency is applied continuously. Modern pulsed FT NMR spectrometers use superconducting magnets (e.g. 300 or 500 MHz) and a short pulse of radiofrequencies is applied to promote the nuclei to the higher-energy

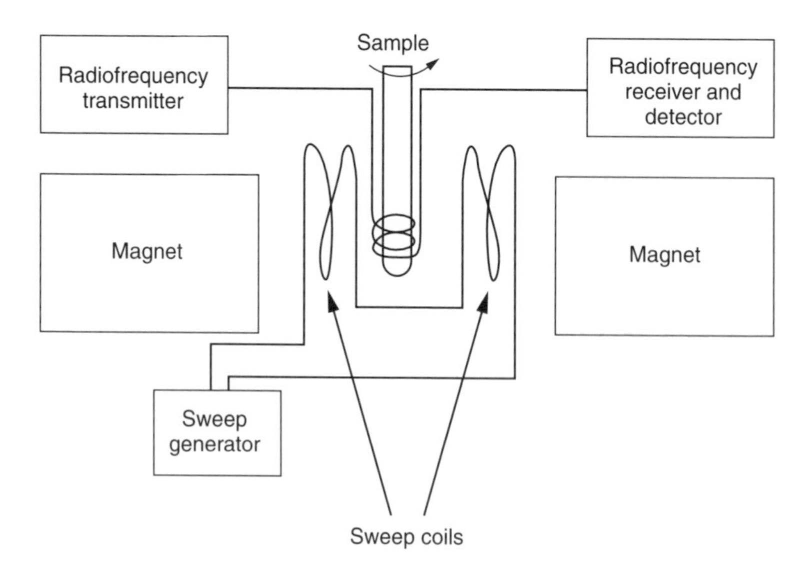

Figure 2.12 Schematic of the layout of a typical NMR spectrometer.

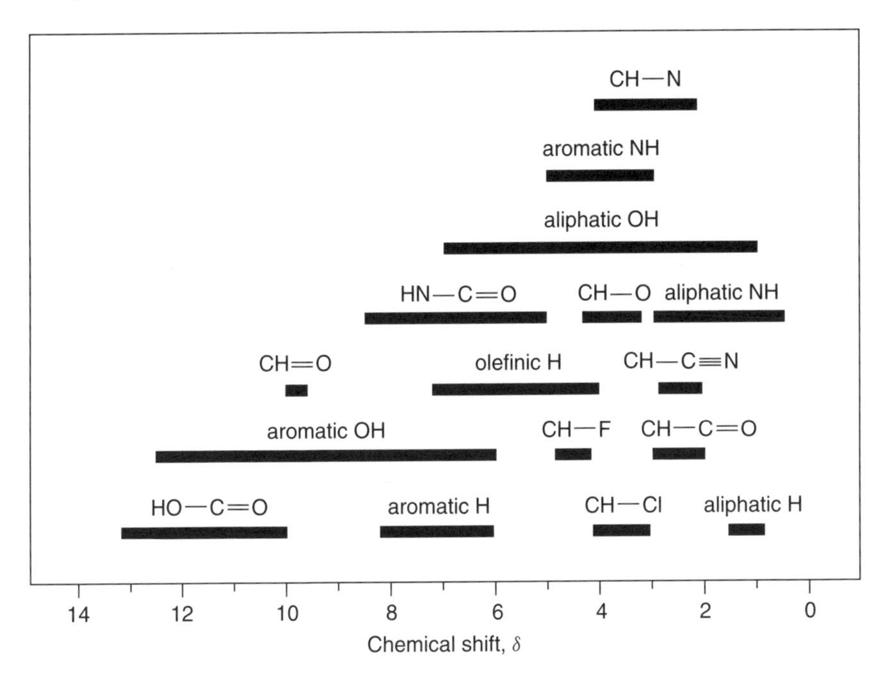

Figure 2.13 A typical correlation table for the ^{1}H NMR chemical shifts of polymers. From R. J. Young and P. A. Lovell, *Introduction to Polymers*, Chapman and Hall, 1991. Reproduced by permission of Kluwer Academic Publishers.

states. The relaxation of the nuclei back to the lower-energy state is detected as an interferogram – known as *free induction decay* (FID) – which can be converted into a spectrum by the process of Fourier transformation.

Some of the parameters of an NMR spectrum that may be used for polymer characterization are chemical shift, peak area, linewidth, coupling constants and relaxation rates. ^{1}H and ^{13}C are the most common nuclei used for the analysis of polymers. Figures 2.13 and 2.14 shows correlation tables for commonly observed functional groups in polymers for ^{1}H and ^{13}C NMR spectroscopies, respectively.

Figure 2.15. shows the ^{13}C NMR spectrum of a crystalline sample of polypropylene (PP), illustrating the distinct chemical shifts corresponding to the CH_2, CH and CH_2 carbons [16], while Figure 2.16 shows the proton-decoupled ^{13}C NMR spectrum of a linear poly(ethylene oxide) (PEO) with hydroxyl end-groups [17]:

$$HO-\overset{a}{C}H_2-\overset{b}{C}H_2-O-\overset{c}{C}H_2-\overset{d}{C}H_2-O\left[\overset{d}{C}H_2-\overset{d}{C}H_2-O\right]\overset{d}{C}H_2-\overset{c}{C}H_2-O-\overset{b}{C}H_2-\overset{a}{C}H_2-OH$$

The main-chain carbon atoms show a strong peak at $\delta = 70.3$ ppm. The weak peaks at 72.5, 70.1 and 61.2 ppm are due to the carbon atoms near the chain ends

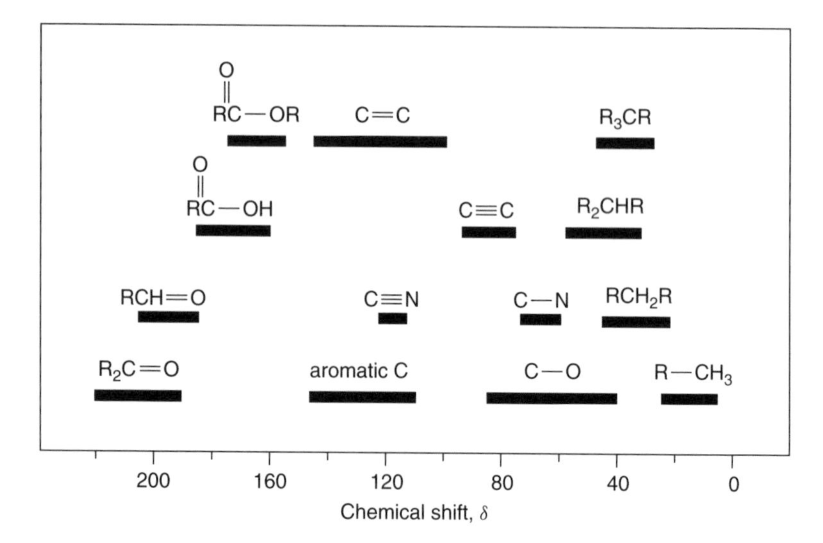

Figure 2.14 A typical correlation table for the ^{13}C NMR chemical shifts of polymers. From R. J. Young and P. A. Lovell, *Introduction to Polymers*, Chapman and Hall, 1991. Reproduced by permission of Kluwer Academic Publishers.

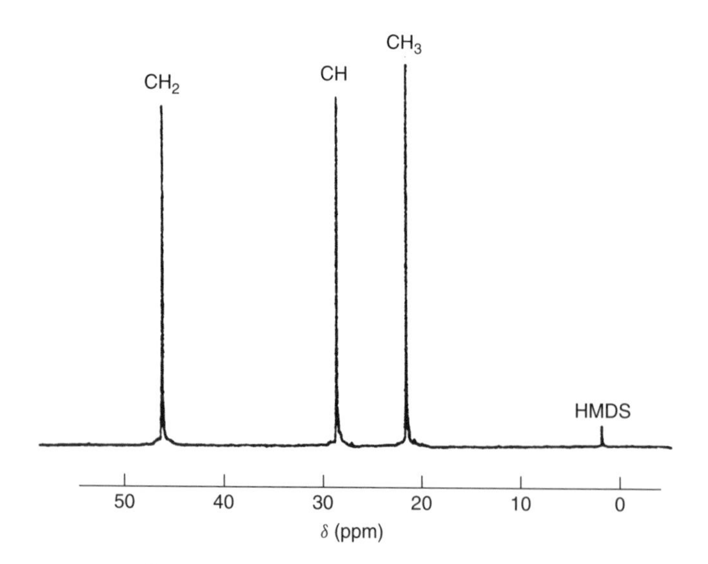

Figure 2.15 The ^{13}C NMR spectrum of a crystalline sample of polypropylene.

and are shifted because of the presence of the hydroxyl end-groups. Note that the molecular weight of such a polymer may be estimated from the ratio of the integrations for the end-group peak to that for the main-chain peak (end-group analysis is discussed further in Chapter 4).

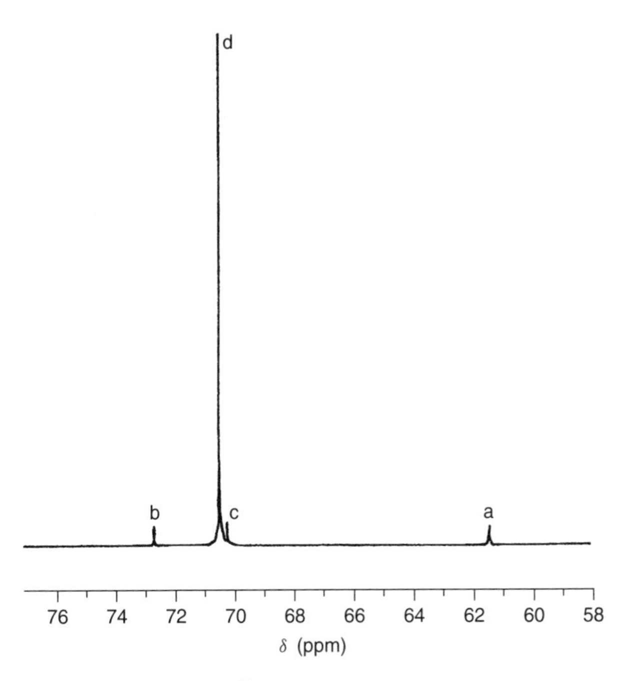

Figure 2.16 The proton-decoupled ^{13}C NMR spectrum of poly(ethylene oxide). From R. J. Young and P. A. Lovell, *Introduction to Polymers*, Chapman and Hall, 1991. Reproduced by permission of Kluwer Academic Publishers.

NMR spectroscopy is commonly used to identify and characterize copolymers. For example, ^{1}H NMR spectroscopy has been used to study copolymers of methyl methacrylate (MMA) and hexyl methacrylate (HMA), with the following structure [18]:

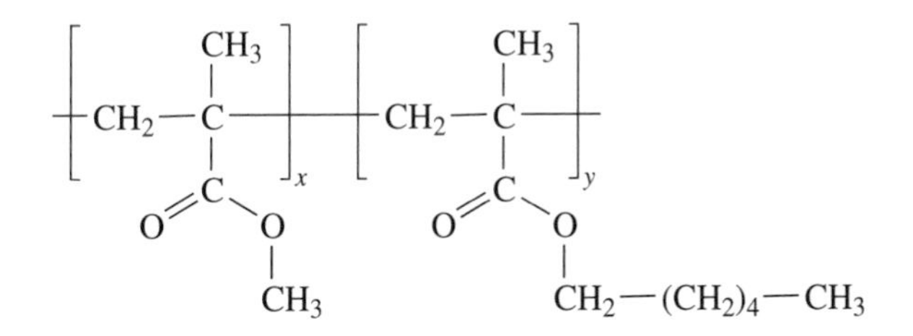

An ^{1}H NMR spectrum of a poly(MMA-*co*-HMA) sample is illustrated in Figure 2.17. The peaks in the range 0.5–2.5 ppm are assigned to alkyl methylene and methyl protons in the copolymer, while the peaks at 3.6 and 3.9 ppm are assigned to the three protons of the OCH_3 group of MMA and the two protons of the OCH_2 group of HMA, respectively. The 3.6 and 3.9 ppm peaks may be used for the quantitative analysis of MMA/HMA copolymers, and the percentage

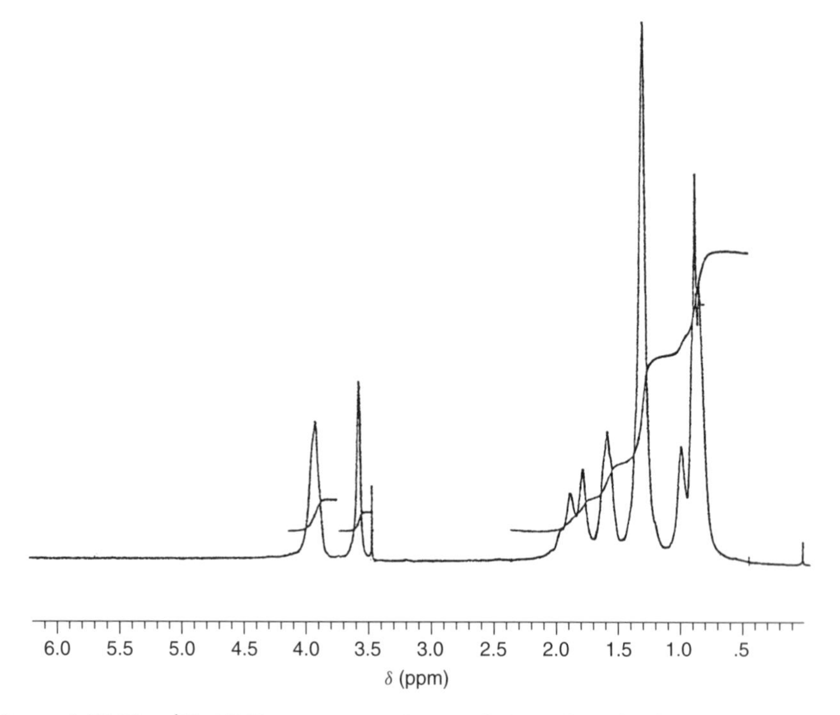

Figure 2.17 The ^{1}H NMR spectrum of a methyl methacrylate/hexyl methacrylate copolymer. Reprinted with permission from P. C. Painter and M. M. Coleman, *Fundamentals of Polymer Science. An Introductory Text*, Technomic Publishing (1997). Copyright CRC Press, Boca Raton, Florida.

of MMA in the copolymer can be calculated by using the following expression:

$$\%\text{MMA} = 100 \times (A_{3.6\ \text{ppm}}/3)/(A_{3.6\ \text{ppm}}/3 + A_{3.9\ \text{ppm}}/2) \quad (2.4)$$

where A is the area under the specific peak.

SAQ 2.7

Determine the composition of the MMA/HMA copolymer shown in Figure 2.17. The areas under the lines at 3.6 and 3.9 ppm are 37 and 61, respectively.

2.6 Ultraviolet–Visible Spectroscopy

The absorption of ultraviolet or visible radiation by polymers leads to transitions among the electronic energy levels and an electronic absorption spectrum results [19]. Ultraviolet–visible (UV–vis) spectroscopy can be used to identify

Table 2.8 Typical chromophores in UV–vis spectroscopy commonly found for polymers

Chromophore	Wavelength, λ_{max} (nm)	Molar absorptivity, ε_{max}
C=C	175	14 000
	185	8000
C≡C	175	10 000
	195	2000
	223	150
C=O	160	18 000
	185	500
	280	15
C=C–C=C	217	20 000
Benzene ring	184	60 000
	200	4400
	255	204

chromophores such as benzene rings and carbonyl groups. It can also be used to determine the lengths of sequences of conjugated multiple bonds in certain polymers. *Chromophores* are functional groups responsible for electronic absorption, with these groups undergoing $n \rightarrow \pi^*$ or $\pi \rightarrow \pi^*$ transitions. Table 2.8 lists some of the chromophores commonly found in polymers [19]. UV–vis spectroscopy is limited to polymers containing appropriate groups, including polymers with aromatic ring or carbonyl groups, and polyenes.

Quantitative analysis in UV–vis spectroscopy can be carried out by using the *Beer–Lambert law*, which is given as follows:

$$A = \log_{10}(I_0/I) = \varepsilon c l \tag{2.5}$$

where A is the absorbance of the solution, I_0 is the intensity of the incident light, I is the intensity of the light transmitted through the sample solution, ε is the molar absorptivity, c is the concentration of the absorbing species, and l is the pathlength of the cell. The molar absorptivity, ε, is a proportionality constant and has a constant value for a particular compound at a given wavelength. Where ε is very large, it is often convenient to express this value as a logarithm ($\log_{10} \varepsilon$). The transmittance (T) is also measured in UV–vis spectroscopy, and is given by the following:

$$T = I/I_0 \tag{2.6}$$

Thus, the relationship between absorbance and transmittance is as follows:

$$A = -\log_{10} T \tag{2.7}$$

The Beer–Lambert law implies that a plot of absorbance against concentration will be linear, with a gradient of εl, and pass through the origin.

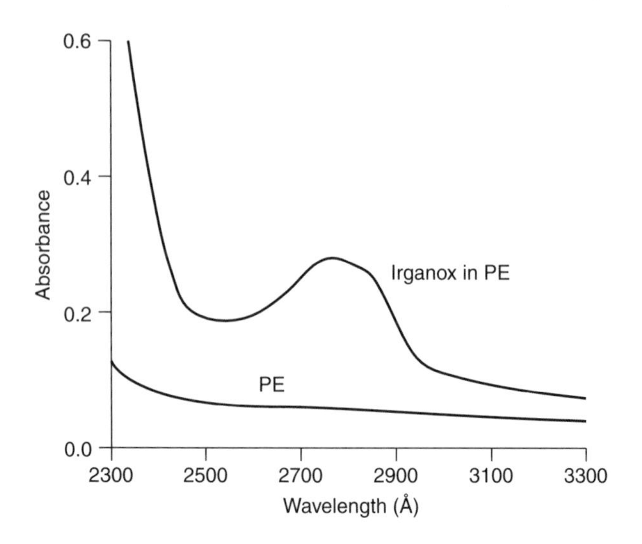

Figure 2.18 The UV–vis spectra of a 'pure' polyethylene sample and one containing the antioxidant Irganox 1010.

Ultraviolet and visible spectra are usually recorded using dilute solutions, and the solvent used must be transparent within the wavelength range being measured. Quartz cells, usually with a pathlength of 1 cm, are employed. A UV–vis spectrum is normally plotted as absorbance or transmittance versus wavelength. The data may also be converted to a plot of molar absorptivity (ε_{max} or $\log_{10} \varepsilon_{max}$) as a function of wavelength. The use of molar absorptivity has the advantage that all of the intensity values refer to the same number of absorbing species.

UV–vis spectroscopy may be employed to identify the presence of any residual monomer remaining in a polymer sample. For example, the presence of styrene in polystyrene may be monitored by using the peak $\lambda_{max} = 292$ nm, which results from the monomer. This approach has also been found to be useful for identifying polymer additives, such as inhibitors and antioxidants. Figure 2.18 illustrates the UV–vis spectrum of polyethylene (PE) containing the 'Irganox 1010' antioxidant (pentaerythritol-tetra-β-(3,5-di-*tert*-butyl-4-hydroxy-phenyl)propionate), as well as the spectrum of PE containing no antioxidant. The antioxidant absorption at 280 nm can be used to carry out quantitative analysis [16]. The composition of copolymers may be determined using UV–vis spectroscopy by measuring the relative areas under suitable absorptions bands.

SAQ 2.8

The benzene ring in polystyrene (PS) absorbs strongly in the ultraviolet region (at 190–200 nm). Butadiene does not absorb in this region. A series of styrene–butadiene copolymers were examined by using UV–vis spectroscopy and the results obtained are summarized in Table 2.9 below. Use this information to determine the composition of the unknown styrene–butadiene copolymer, which records an absorbance value of 0.349 at 195 nm under the same conditions.

Table 2.9 UV–vis spectroscopic data obtained for a series of styrene–butadiene copolymers (SAQ 2.8)

Styrene content (%)	Absorbance at $\lambda_{max} = 195$ nm
5	0.103
10	0.205
15	0.308
20	0.410
25	0.513

2.7 Differential Scanning Calorimetry

The term *thermal analysis* describes a number of techniques which involve measuring a physical property of a material as a function of temperature [20]. *Differential scanning calorimetry* (DSC) and *differential thermal analysis* (DTA) are two such thermal methods that may be used to compare and evaluate polymer samples. DSC is a technique which records the energy necessary to establish a zero temperature difference between the sample and a reference material as a function of time or temperature. In this method, the two specimens are subjected to identical temperature conditions in an environment which is heated or cooled at a controlled rate. In contrast, DTA involves measuring the difference in temperature between the sample and the reference material as a function of time or temperature. A schematic diagram showing the layout of the experimental set-up used to carry out DSC or DTA is shown in Figure 2.19.

DSC curves are plotted as a function of time or temperature at a constant rate of heating. Figure 2.20 shows a typical curve obtained for a polymer sample. A shift in the baseline results from the change in heat capacity of the sample. The basic equation used in DSC analysis is as follows:

$$\Delta T = qC_p/K \tag{2.8}$$

where ΔT is the difference in temperature between the reference material and the sample, q is the heating rate, C_p is the heat capacity (at constant pressure), and K is a calibration factor for the particular instrument being used.

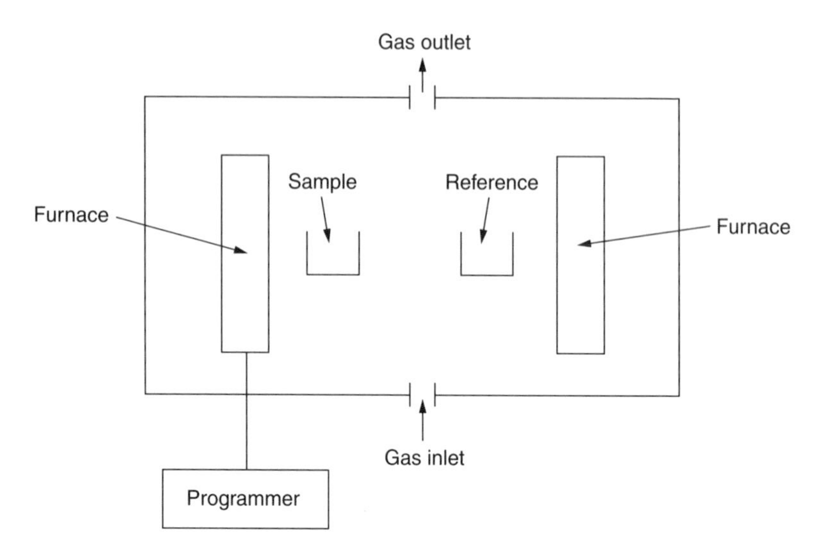

Figure 2.19 Schematic of the experimental set-up used to carry out differential scanning calorimetry or differential thermal analysis.

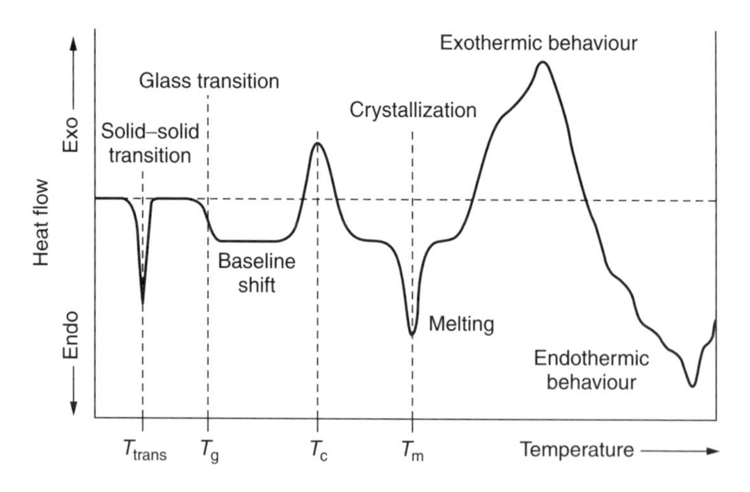

Figure 2.20 A typical DSC curve obtained for a polymer sample.

DSC can be used for measuring the enthalpy involved in polymer transitions. The peak area between the curve and the baseline is proportional to the enthalpy change (ΔH) in the sample. This enthalpy change can be determined from the area of the curve peak (A) by using the following relationship:

$$\Delta Hm = KA \tag{2.9}$$

where m is the mass of the polymer sample and K is a calibration coefficient dependent on the instrument being employed for the measurements.

Thermal methods can be used to identify polymers by comparison with standard reference curves. For example, Figure 2.21 shows the DSC traces obtained for nylon 6 and nylon 6,10. Transitions due to specific thermal processes can be identified – these will be examined in further detail in Chapter 5. Transitions can

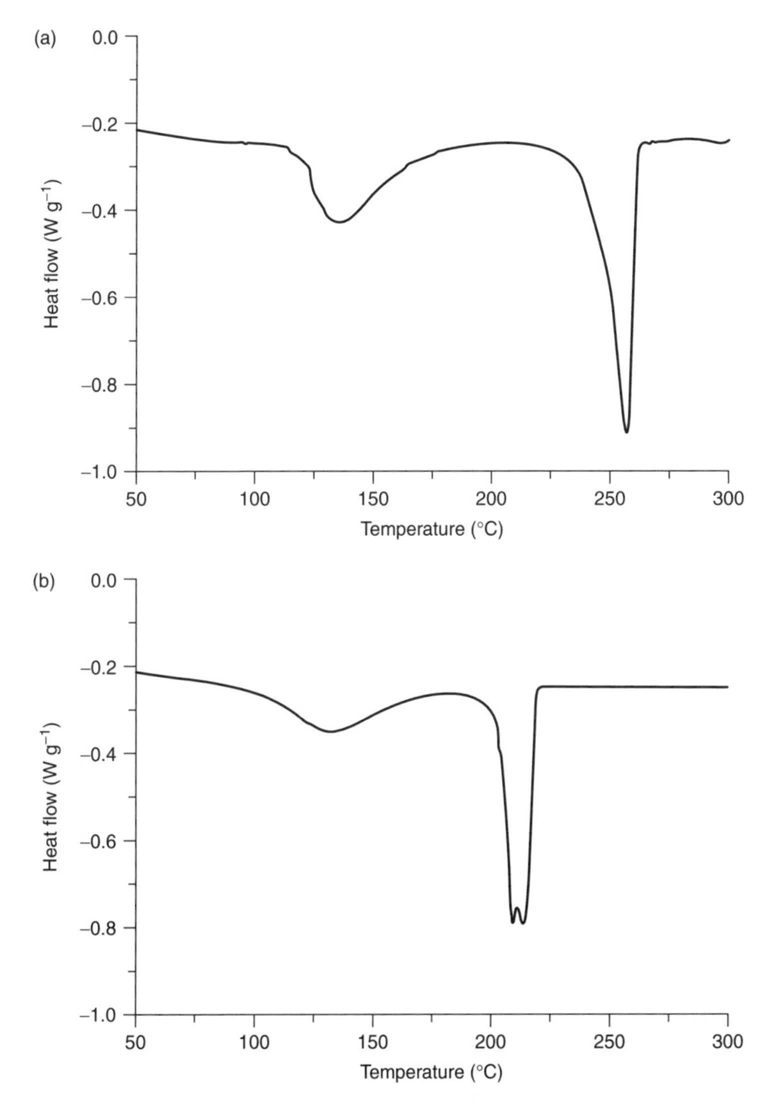

Figure 2.21 DSC traces obtained for samples of (a) nylon 6 and (b) nylon 6,10.

also appear in DSC due to chemical reactions such as polymerization, curing, oxidation or cross-linking (see Chapter 3).

SAQ 2.9.

Many polymer bottles used for household chemicals and products such as shampoos are made of recycled high-density polyethylene (HDPE). The major impurity present in such recycled polymer would be the bottle lids which are produced from polypropylene (PP). The DSC trace of a 10.5 mg sample of this type of polymer is shown below in Figure 2.22, while typical thermal analysis data for HDPE and PP are presented below in Table 2.10 [20]. Determine the enthalpy of each peak (in J g^{-1}) of the (polymer) mixture, plus estimate the percentage amount of PP in the sample.

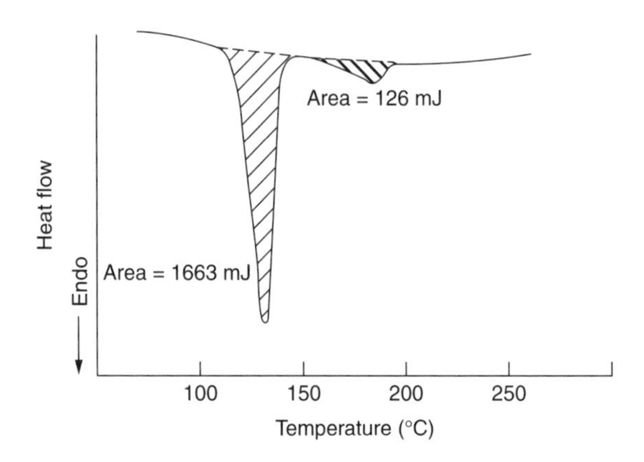

Figure 2.22 DSC trace obtained for a sample of recycled polymer containing high-density polyethylene and polypropylene (SAQ 2.9).

Table 2.10 Thermal analysis data obtained for polypropylene and high-density polyethylene (SAQ 2.9)

Polymer	T_g (°C)	T_m (°C)	ΔH_m (J g^{-1})
PP	−20	17	100
HDPE	−120	13	18

2.8 Mass Spectrometry

Mass spectrometry (MS) involves the study of ions in the gas phase [21]. A mass spectrum represents the ion abundance versus the mass-to-charge (m/z) ratio of the ions separated in a mass spectrometer. Figure 2.23 illustrates the general

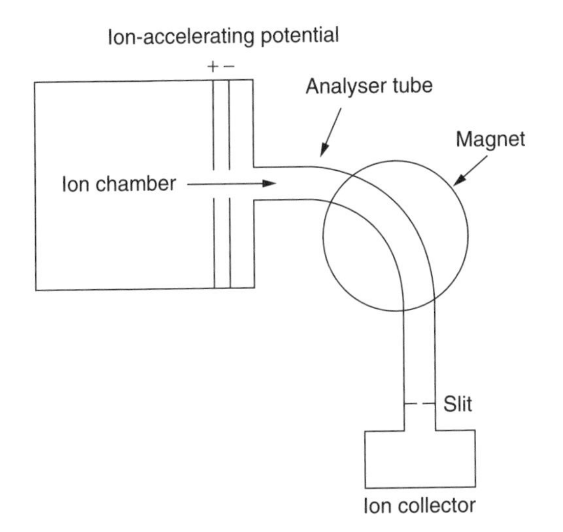

Figure 2.23 Schematic of the layout of a typical mass spectrometer.

layout of a typical mass spectrometer. In this set-up, the ion source vaporizes and ionizes the sample – common ionization methods include electron impact and chemical ionization. In electron impact, the molecules are volatalized by heat and ionized with a beam of electrons from a hot wire filament. The most probable event involves the ejection of an electron, thus resulting in a cation radical $[M]^{+\bullet}$ of the same mass as the initial molecule M, as follows:

$$M + e^- \longrightarrow [M]^{+\bullet} + 2e^- \tag{2.10}$$

The chemical ionization technique has proved to be more successful for the analysis of polymers. This method involves the introduction of an intermediate substance (e.g. methane) at a higher concentration than that of the sample under investigation. The carrier gas is ionized by electron impact and the sample is then ionized by collisions with these ions. Chemical ionization has a 'softer' ion source than electron impact and so produces less fragmentation.

More recently introduced techniques, such as *matrix-assisted laser desorption ionization* (MALDI) and *fast-atom bombardment* (FAB), involve the combination of ion sources with desorption ionization, where vaporization and ionization occur essentially at the same time. This approach is useful for non-volatile and high-molecular-weight molecules. The mass analyser separates ions according to the m/z ratio. A *time-of-flight* (TOF) mass analyser measures the flight time of ions with an equivalent kinetic energy over a fixed distance.

MS is a widely used technique for the identification of materials and characteristic mass spectra for many compounds have been collected in databases.

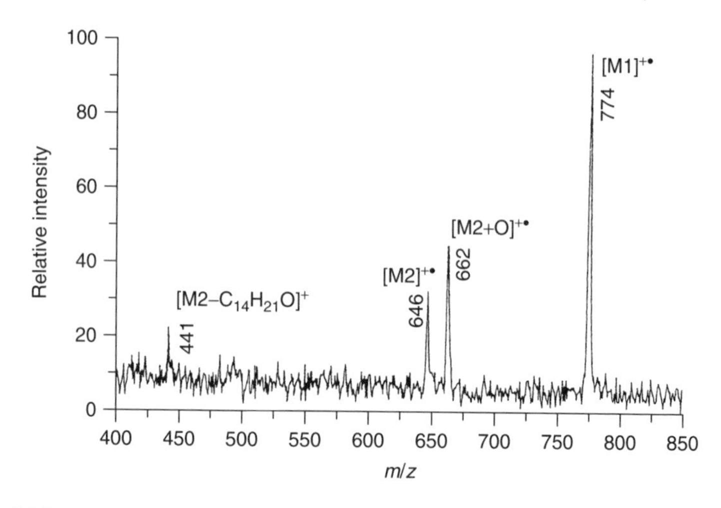

Figure 2.24 Laser-desorption mass spectrum of a polypropylene sample containing Irganox 1330 (0.15 wt%) and Irganox 168 (0.05 wt%); $[M1]^{+\bullet}$ and $[M2]^{+\bullet}$ represent molecular ions of Irganox 1330 and Irganox 168, respectively. Reprinted with permission from Wright, S. J., Dale, M. J., Langridgesmith, P. R. R., Zhan, Q. and Zenobi, R., *Anal. Chem.*, **68**, 3585–3594 (1996). Copyright (1996) American Chemical Society.

Identification relies on the fact that the fragmentation of molecules is reproducible. For example, the mass spectrum of a copolymer allows the average composition to be determined.

MS has proved particularly useful for the identification of polymer additives. This method avoids the need for extraction and is able to detect low concentrations of additives. Figure 2.24 shows the laser-desorption mass spectrum of a polypropylene (PP) sample containing the additives 'Irganox 1330' (1,3,5-tris(3,5-di-*tert*-butyl-4-hydroxybenzyl)-2,4,6-trimethylbenzene) (0.15 wt%) and 'Irgafos 168' (tris(2,4-di-*tert*-butylphenyl)phosphite) (0.05 wt%) [22]. The molecular ions for both Irganox 1330 ($m/z = 774$) and Irgafos 168 ($m/z = 646$) are observed. The peak at $m/z = 662$ is attributed to the oxide of Irgafos 168, while the less intense peak at $m/z = 441$ corresponds to the loss of a 2,4-di-*tert*-butylphenyl-O side-group in Irgafos 168 via direct cleavage.

2.9 Chromatography

A number of chromatographic techniques are suitable for polymer analysis [23]. These methods all involve the separation of sample components. Gas chromatography, high performance liquid chromatography and thin layer chromatography are all techniques that may be used for identification purposes and will be discussed in this section.

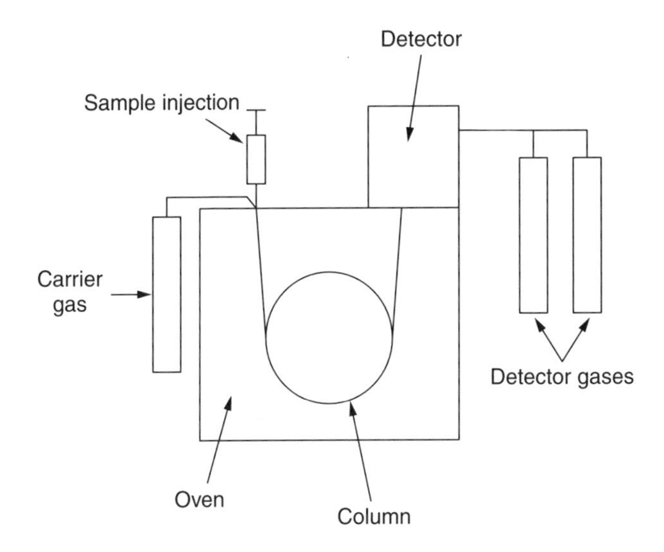

Figure 2.25 Schematic of the layout of a typical gas chromatograph.

Gas chromatography (GC) can be used to study the volatile components of polymers. Figure 2.25 illustrates the layout of a typical gas chromatograph. In GC, the sample in a gaseous mobile phase is passed through a column containing the liquid stationary phase. The retention (of the sample) depends on the degree of interaction with the liquid phase and its volatility. The column is usually heated in the range 50–300°C, with a flame-ionization detector commonly being employed as the detection system.

Standard GC can be used to quantify polymer additives and monomers. This technique has been widely used to monitor volatiles in polymers used, for instance, in packaging and paints. The strength of GC has been the ability to combine this technique with other analytical instrumentation. For example, gas chromatography can be combined with mass spectrometry and infrared spectroscopy to identify the separated components. The technique of pyrolysis GC–MS involves studying the degradation products of polymers and is discussed later in Chapter 7. Inverse gas chromatography can be used to study the surface properties of polymers and is discussed below in Chapter 6.

High performance liquid chromatography (HPLC) is used to identify and quantify the low-molecular-weight species present in polymers. Figure 2.26 shows a schematic diagram of a typical HPLC instrument. The technique involves injecting the sample into the solvent stream (mobile phase), where the latter usually consists of two or more solvents that may be varied in composition with time. The sample flows through a column packed with silica and the separate bands are then detected by using an ultraviolet or refractive index detector. There are two separation modes used for polymer analysis, i.e. normal phase and reverse

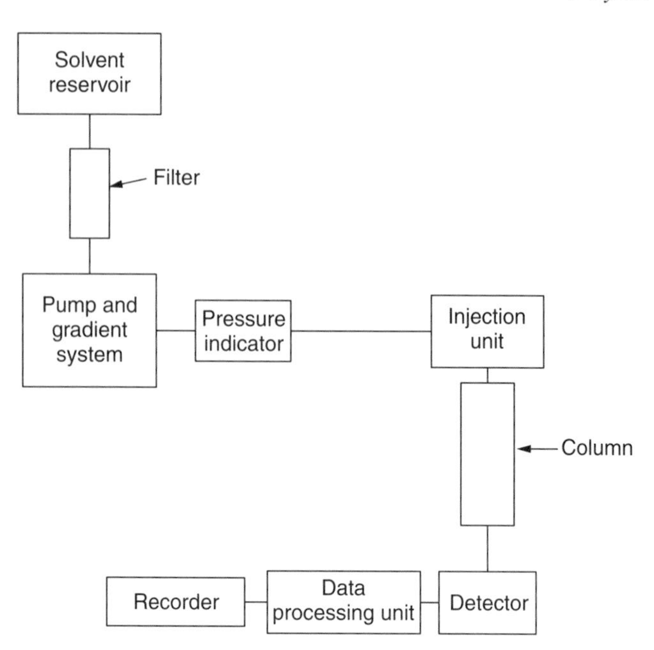

Figure 2.26 Schematic of the layout of a typical high performance liquid chromatograph.

phase. Normal-phase chromatography involves separations carried out by using a polar stationary phase (e.g. silica or bonded silica) and a non-polar organic mobile phase (e.g. hydrocarbons). In reverse-phase chromatography, the stationary phase is non-polar (e.g. octadecyl silane), with the mobile-phase being polar in nature (e.g. water, acetonitrile or methanol).

HPLC has been widely used for the quantitative analysis of additives in polymers. Additives such as antioxidants, stabilizers and plasticizers may be determined in this way. Higher-molecular-weight constituents, such as resin components and oligomers (containing a series of only a few monomer units), can also be separated. In addition, it is possible to analyse the composition of various copolymers. As an example of such an application, Figure 2.27 shows the HPLC analysis data obtained for various PS/PMMA copolymers on a silica column [24]. The chromatogram resulting from a mixture of seven such copolymers, ranging in composition from 11.4 to 88.5 wt% MMA, is illustrated by the continuous line and shows clearly the different elution times of each component. This information may be used to characterize copolymers of unknown composition. For example, the dashed line given in Figure 2.27 represents the trace obtained from an 'unknown' PS/PMMA copolymer – a simple calculation reveals that the latter contains 73.8 wt% MMA.

Thin layer chromatography (TLC) is a very simple chromatographic method which may also be used to identify polymers. In this technique, the stationary

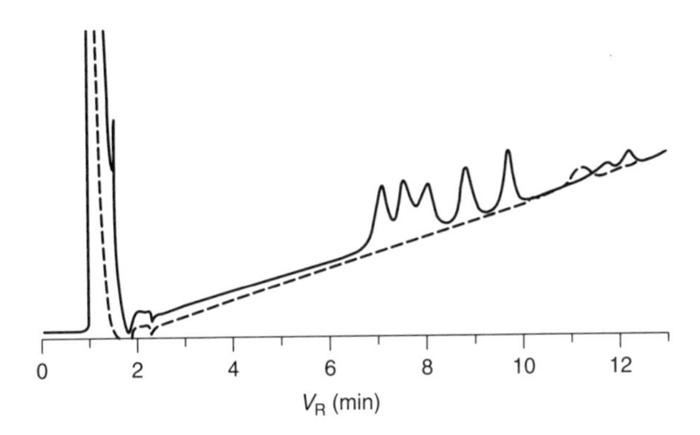

Figure 2.27 HPLC analysis of PS/PMMA copolymers: (——) a mixture of seven copolymers ranging in composition from 11.4 to 88.5 wt% MMA; (- - - -) a single copolymer sample containing 73.8 wt% MMA. From G. Glöckner and A. H. E. Müller, *J. Appl. Polym. Sci.* **38**, 1761–1774 (1989). Reproduced by permission of John Wiley & Sons, Inc.

SAQ 2.10

HPLC analysis was carried out on a series of PS/PMMA copolymers, using a chloroform/ethanol solvent system, on a silica column at 20°C. Figure 2.28 below shows the HPLC retention volumes determined for these copolymers as a function of their compositions. A similar copolymer, of unknown composition, was then applied to the column under the same experimental conditions and showed a retention volume of 5.5 ml. Estimate the composition of this material.

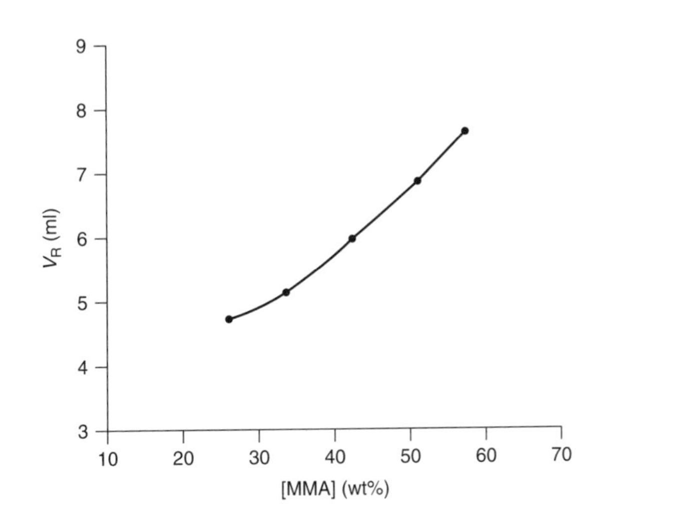

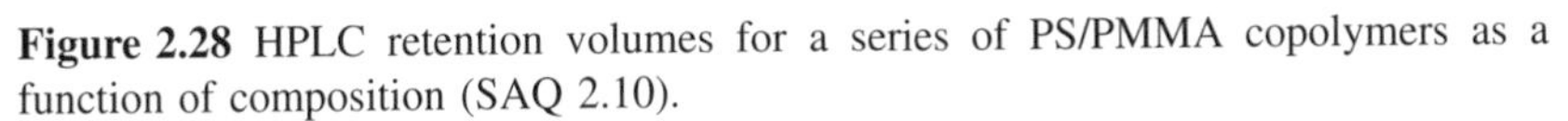

Figure 2.28 HPLC retention volumes for a series of PS/PMMA copolymers as a function of composition (SAQ 2.10).

phase (e.g. silica) is coated as a thin layer on to a plate. The polymer (in solution) is applied as a spot on the bottom of the plate, and the latter is then placed in a tank containing the mobile phase, i.e. a mixture of solvents. The separation occurs as the solvent is carried up the plate by capillary action. The separated components are usually detected by spraying the plate with a suitable visualizing agent. TLC can be used to identify and characterize homopolymers, copolymers and additives.

2.10 Emission Spectroscopy

When molecules absorb radiation in electronic transitions to form excited states, the latter may lose the acquired energy via several mechanisms. If the energy loss occurs through the emission of radiation, the process is known as *luminescence* [19]. *Fluorescence* is a form of luminescence and involves emission occurring from the lowest excited single state (S_1) to the singlet ground state (S_0), as illustrated in Figure 2.29. Fluorescence ceases immediately the exciting radiation is removed, with the lifetime usually being of the order of 10^{-8}–10^{-10} s. The orange and green colours of some of the fluorescent dyes in common use are examples of this phenomenon – such dyes absorb in the ultraviolet and blue electromagnetic regions and fluoresce in the visible region. *Phosphorescence* is

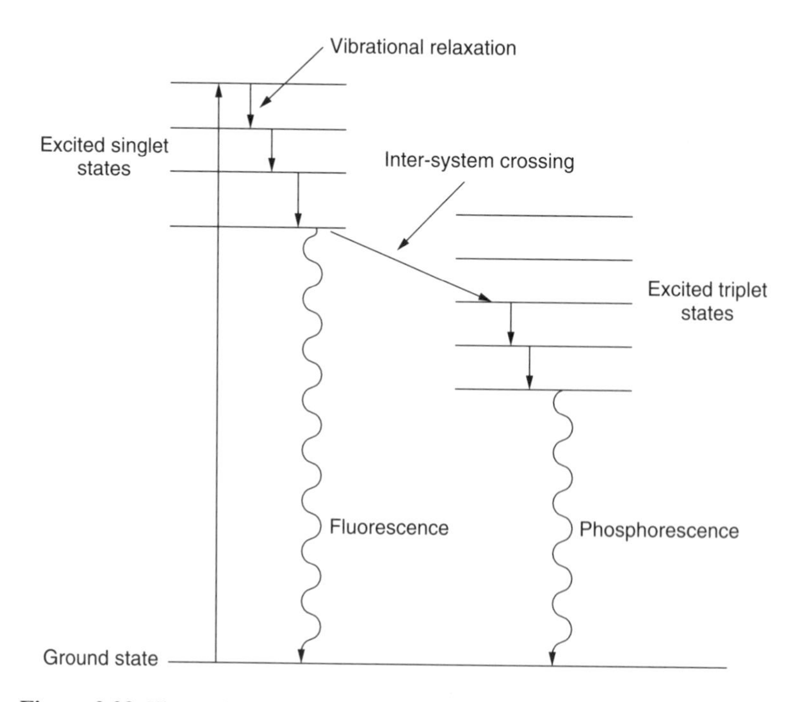

Figure 2.29 Illustration of the processes of fluorescence and phosphorescence.

another from of luminescence and involves emission occurring from the lowest excited triplet state (T_1) to the singlet ground state (S_0) (see Figure 2.29). Such a process involves inter-system crossing, with the lifetimes in this case ranging from 10^{-3} s up to several minutes.

An *excimer* is a dimer aggregate formed between an excited molecule in the S_1 state and a molecule in the S_0 state. Excimers are unstable in their ground states, but are stable under electronic excitation. During the decomposition of an excimer complex, *excimer fluorescence* may be observed. Excimers may also result from the reactions of two T_1 states. During the decomposition of such an excimer, *delayed fluorescence* may be observed, which is an emission that has the spectral properties of fluorescence, but with a longer rise and decay time than normal fluorescence.

Fluorescence and phosphorescence spectra are usually measured by viewing the emission at angles of 90° to the direction of the excitation. In the instrument used for this purpose, which is, known as a *fluorimeter*, the exciting light is selected in a monochromator and a second monochromator is then used to scan the emitted light from near to the excitation wavelength up to longer wavelengths. Phosphorescence is more difficult to observe than fluorescence because of the possibility of quenching by impurities that may be present in a sample.

The fluorescence and excitation spectra of a specific molecule are often mirror images of one another. A fluorescence spectrum is a plot of the fluorescence intensity (I_f) as a function of the wavelength. Figure 2.30 illustrates the general form of such a spectrum. If the I_f value can be determined by using the Beer–Lambert law, then we have the following relationship:

$$I_f = I_0(1 - 10^{-\varepsilon cl})q \tag{2.11}$$

where I_0 is the intensity of the incident light, ε is the molar absorptivity, c is the concentration of the solute (mol l^{-1}), l is the sample pathlength (cm) and q is the quantum yield of fluorescence. The latter parameter is the number of quanta emitted per exciting quantum absorbed. Unfortunately, q is a difficult quantity to measure. However, the relative fluorescence yield can be determined when it is possible to make comparisons with standards which have known quantum yields.

A phosphorescence spectrum is a plot of phosphorescence intensity (I_P) as a function of the wavelength (see Figure 2.30). The phosphorescence intensity decays after the excitation source is removed, according to the following relationship:

$$dI_P/dt = k_1 I_P + k_2 I_P^2 \tag{2.12}$$

where k_1 and k_2 are rate constants. Phosphorescence lifetime measurements can be made using fluorimeters by employing a 'chopped-light source' (known as the *flash method*).

Emission spectra can be used for the identification of polymers. Tables 2.11 and 2.12 summarize, respectively, the fluorescence and phosphorescence characteristics of some common polymers [25]. Emission spectra can also be used

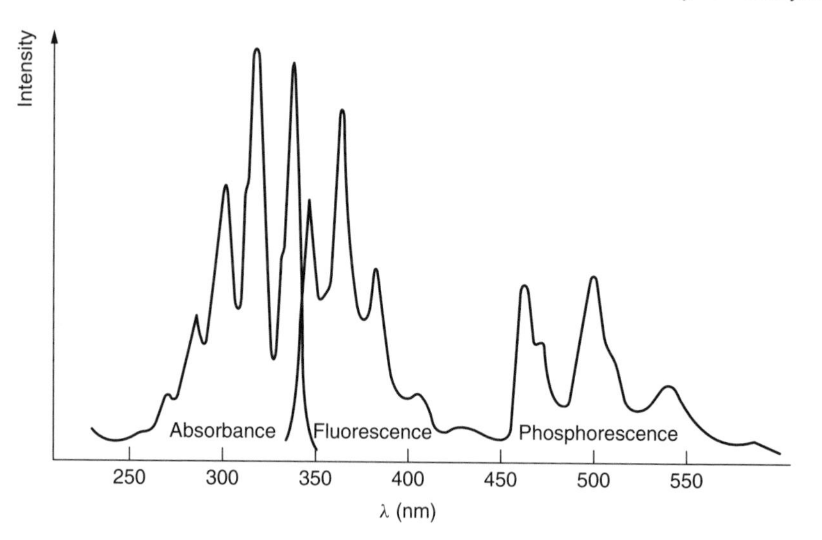

Figure 2.30 A spectrum showing the general features of both fluorescence and phosphorescence.

Table 2.11 Fluorescence characteristics of some common polymers

Polymer	Excitation (nm)	Emission (nm)
PU	372	420
Nylon 6,6	357	417
Nylon 6	335	390
Nylon 6,10	345, 355	395, 410
Nylon 11	327, 340	375, 385, 395
Nylon 12	410	450
PVF	290	350
PTFE	328	350
PVAl	258, 295, 330	360, 370
PBT	332	400, 420, 450
LDPE	230, 273	295, 310, 329, 354, 370
HDPE	230, 265, 290	295, 312, 330, 344, 358
PP	230, 285	309, 320
PS	318, 330	336, 354, 368
PES	320	360
Epoxy resin	350	424

Table 2.12 Phosphorescence characteristics of some common polymers

Polymer	Excitation (nm)	Emission (nm)	Lifetime (s)
PET	280, 318, 351	425, 460	0.5
PU	320	423, 455, 489	0.02
Nylon 6,6	296	400, 430	1.3–2.1
Nylon 6	282	390, 420, 455	1.1–1.7
Nylon 6,10	300	430	0.7
Nylon 11	296, 273	423, 450	0.88–1.0
Nylon 12	268, 286	363, 410	1.0
PVC	284	440	0.3
PTFE	260–280	450	0.4
PVAl	260–280	436	0.4
PBT	305	450	1.2
LDPE	273, 280, 278, 283, 331	367, 381, 391, 405, 416, 420, 370, 335, 455	0.6–2.3
HDPE	275	450	0.35
PP	270, 290, 330	420, 445, 480, 510	0.5–1.2
PS	290, 300, 336	398, 425, 456, 492	0.008
PES	320	450	0.05
Epoxy resin	305	460	–

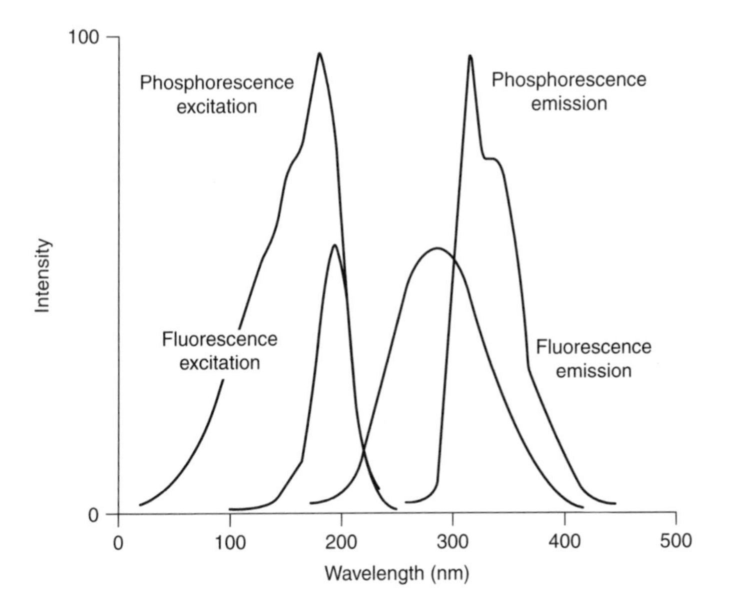

Figure 2.31 The fluorescence and phosphorescence spectra of the polymer stabilizer Nonex CI.

to detect the presence of additives in polymers. In particular, the fluorescence and phosphorescence spectra of antioxidants and stabilizers have been used for this type of analysis [26]. In addition, pigments such as dyes exhibit their own characteristic emission spectra. For example, different crystalline forms of titanium dioxide may be differentiated by their different characteristic emissions. Figure 2.31 demonstrates the application of emission spectroscopy to additive analysis, illustrating the fluorescence and phosphorescence spectra observed for the stabilizer 'Nonox CI' (N,N'-di-β-naphthyl-p-phenylenediamine) [26]. This stabilizer displays fluorescence excitation and emission peaks at 392 and 490 nm, respectively, while its phosphorescence spectrum shows excitation and emission peaks, respectively, at 382 and 516 nm. The phosphorescence lifetime is 0.90 s.

Summary

This chapter has described the experimental techniques that are commonly used to identify polymers and their additives. Preliminary identification methods, such as solubility, density, flame tests, pyrolysis tests and melting temperature, were introduced. Standard analytical methods, including infrared spectroscopy, Raman spectroscopy, nuclear magnetic resonance spectroscopy, ultraviolet–visible spectroscopy, differential scanning calorimetry, mass spectrometry, chromatographic techniques and emission spectroscopy, were then described and how these techniques may be applied to polymeric systems was outlined.

References

1. Braun, D., *Simple Methods for Identification of Plastics*, Hanser, Munich, 1996.
2. Schroder, E., Muller, G. and Arndt, K. F., *Polymer Characterization*, Hanser, Munich, 1989.
3. Bower, D. I. and Maddams, W. F., *The Vibrational Spectroscopy of Polymers*, Cambridge University Press, Cambridge, UK, 1989.
4. Siesler, H. W. and Holland-Moritz, K., *Infrared and Raman Spectroscopy of Polymers*, Marcel Dekker, New York, 1980.
5. Pouchert, C. J., *The Aldrich Library of FTIR Spectra*, Aldrich Chemical Company, Milwaukee, WI, 1989.
6. Hendra, P. J., Jones, C. and Warnes, G., *Fourier Transform Raman Spectroscopy: Instrumentation and Chemical Applications*, Ellis Horwood, New York, 1991.
7. Agbenyega, J. K., Ellis, G., Hendra, P. J., Maddams, W. F., Passingham, C., Willis, H. A. and Chalmers, J., *Spectrochim. Acta, A*, **46**, 197–216 (1990).
8. Maddams, W. F., *Spectrochim. Acta, A*, **50**, 1967–1986 (1994).
9. Xue, G., *Prog. Polym. Sci.*, **22**, 313–406 (1997).
10. Hendra, P. J., Maddams, W. F., Royaud, I. A. M., Willis, H. A. and Zichy, V., *Spectrochim. Acta, A*, **46**, 747–753 (1990).
11. Kupstov, A. H. and Zhizhin, G. N., *Handbook of Fourier Transform Raman and Infrared Spectra of Polymers*, Elsevier, Amsterdam, The Netherlands, 1998.
12. Bovey, F. A., 'Structure of Chains by Solution NMR Spectroscopy', in *Comprehensive Polymer Science*, Vol. 1, Booth, C. and Price, C. (Eds), Pergamon Press, Oxford, UK, 1989, pp. 339–376.
13. McBrierty, V. J., 'NMR Spectroscopy of Polymers in the Solid State', in *Comprehensive Polymer Science*, Vol. 1, Booth, C. and Price, C. (Eds), Pergamon Press, Oxford UK, 1989, pp. 397–428.

14. Mirau, P. A., 'NMR Characterization of Polymers', in *Polymer Characterization*, Hunt, B. J. and James M. I. (Eds), Blackie, London, 1993, pp. 37–68.
15. Sandler, S. R., Karo, W., Bonesteel, J. and Pearce, E. M., *Polymer Synthesis and Characterization: A Laboratory Manual*, Academic Press, San Diego, CA, 1998.
16. Crompton, T. K., *Analysis of Polymers: An Introduction*, Pergamon Press, Oxford, UK, 1989.
17. Young, R. J. and Lovell, P. A., *Introduction to Polymers*, Chapman and Hall, London, 1991.
18. Painter, P. C. and Coleman, M. M., *Fundamentals of Polymer Science: An Introductory Text*, Technomic Publishing, Lancaster, PA, 1997.
19. Rabek, J. F., *Experimental Methods in Polymer Chemistry*, Wiley, Chichester, 1980.
20. Haines, P. J., *Thermal Methods of Analysis: Principles, Applications and Problems*, Blackie, London, 1995.
21. Koenig, J. L., *Spectroscopy of Polymers*, Elsevier, Amsterdam, The Netherlands, 1999.
22. Wright, S. J., Dale, M. J., Langridgesmith, P. R. R., Zhan, Q. and Zenobi, R., *Anal. Chem.*, **68**, 3585–3594 (1996).
23. Handley, A., 'Chromatographic Methods', in *Polymer Characterization*, Hunt, B. J. and James M. I. (Eds), Blackie London, 1993, pp. 145–177.
24. Glockner, G. and Muller, A. H. E., *J. Appl. Polym. Sci.*, **38**, 1761–1774 (1989).
25. Beddard, G. S. and Allen, N. S., 'Emission Spectroscopy', in *Comprehensive Polymer Science*, Vol. 1, Booth, C. and Price, C. (Eds), Pergamon Press, Oxford, UK, 1989, pp. 499–516.
26. Kirkbright, G. F., Narayangswarmy, R. and West, T. S., *Anal. Chim. Acta*, **52**, 237–246 (1970).

Chapter 3

Polymerization

Learning Objectives

- To understand the process of chain polymerization.
- To understand the process of step polymerization.
- To understand the statistics and the kinetics associated with step polymerization processes.
- To understand the process of copolymerization.
- To understand the process of cross-linking of polymers.
- To use dilatometry, infrared spectroscopy, Raman spectroscopy, nuclear magnetic resonance spectroscopy, electron spin resonance spectroscopy and refractometry to monitor polymerization processes.

3.1 Introduction

Polymerization reactions were originally classified as being either *condensation* or *addition* processes, as suggested by W.H. Carothers in 1929. Condensation reactions involve the reaction of monomers accompanied by the loss of a small molecule, such as water, while addition reactions involve the addition of a monomer without such loss. However, there are polymerization reactions that do not fall readily into these classifications, such as those involved in the production of polyurethanes. A clearer classification was developed by P.J. Flory in the 1950s which involved the actual *mechanisms* of the polymerization processes which occurred. Flory proposed the ideas of *chain polymerization* and *step polymerization* mechanisms. We shall describe the features of such mechanisms in this present chapter. As mentioned earlier in Chapter 1, copolymers provide an opportunity to improve the properties of the component homopolymers. Methods for copolymerization cross-linking are also discussed in this chapter. Certain experimental methods may be utilized to monitor the changes undergone as polymerization takes place. These include dilatometry, infrared spectroscopy, Raman spectroscopy, nuclear magnetic resonance spectroscopy, electron spin resonance spectroscopy and refractometry, and these methods will also be described in the following sections.

3.2 Chain Polymerization

During a chain polymerization reaction, the active centre responsible for the growth of the polymer chain is associated with the addition of consecutive monomer units to a single molecule [1, 2]. The characteristic features of chain polymerization are as follows:

- Growth is by the addition of the monomer at one end of the chain.
- Even at long reaction times, some monomer remains in the reaction mixture.
- The molecular weight of the polymer increases rapidly.
- Different mechanisms operate at different stages of the reaction.
- The polymerization rate initially increases and then becomes constant.
- An initiator is required to start the reaction.

Typically, a chain polymerization reaction consists of three stages, namely initiation, propagation and termination. There are a variety of different chain polymerization mechanisms, including free radical, ionic, complex and ring opening.

3.2.1 Free-Radical Chain Polymerization

Free-radical chain polymerization is the most important (chain) polymerization mechanism, usually involving vinyl monomers ($CH_2{=}CHX$ or $CH_2{=}CXY$). During the initiation step, a small trace of *initiator* is required. The latter are materials that readily fragment when heated or irradiated. Benzoyl peroxide and azobisisobutyronitrile (AIBN) are two commonly employed initiators, with the fragmentation processes of these substances being illustrated are follows:

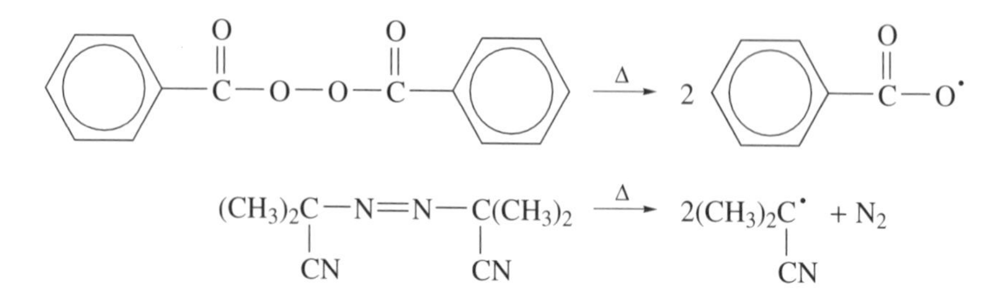

The free radical ($R^\bullet$) produced by the fragmented initiator reacts rapidly with the monomer to yield a new free radical species as follows:

$$R^\bullet + CH_2{=}CHX \longrightarrow RCH_2C^\bullet HX$$

Propagation involves a series of reactions in which the free radical at the end of the growing polymer reacts with the monomer to increase the length of the chain:

$$RCH_2C^\bullet HX + CH_2{=}CHX \longrightarrow RCH_2CHXCH_2C^\bullet HX$$

Polymerization will not proceed until all of the monomer is used up because the free radicals are so reactive. There are two methods of termination, i.e. *combination* and *disproportionation*. During termination by combination, two radical species react together to form a product with a single bond and one reaction product:

$$\sim CH_2C^\bullet HX + \sim CH_2C^\bullet HX \longrightarrow \sim CH_2CHXCHXCH_2 \sim$$

During termination by disproportionation, two radical species react via hydrogen abstraction to form two reaction products, according to the following:

$$\sim CH_2C^\bullet HX + \sim CH_2C^\bullet HX \longrightarrow \sim CH_2CH_2X + \sim CH{=}CHX$$

SAQ 3.1

Vinyl chloride may be polymerized by using benzoyl peroxide as the initiator. Write down a reaction scheme for this process. Assume that termination takes place by combination.

A study of the kinetics of free-radical polymerization allows the rate of propagation to be determined. The equations representing initiation are as follows:

$$I \xrightarrow{k_d} 2R^\bullet$$

$$R^\bullet + M \xrightarrow{k_a} RM^\bullet$$

where k_d and k_a are rate constants. The decomposition of initiator into free radicals is the rate-determining decomposition. Thus, the rate of initiation is given by the following:

$$R_i = d[M^\bullet]/dt = 2fk_d[I] \qquad (3.1)$$

where R_i is the rate of initiation, $[M^\bullet]$ is the radical species concentration, [I] is the initiator concentration, and f is the *initiator efficiency*, which is the fraction of initiator fragments that actually brings about the polymerization. The value of f is typically 0.6–1.0, as newly generated free radicals can recombine before they have time to move apart (this is known as the *cage effect*).

Two general reactions may be written for the termination steps, as follows:

$$M_n^\bullet + M_m^\bullet \xrightarrow{k_{tc}} M_{m+n}$$

$$M_n^\bullet + M_m^\bullet \xrightarrow{k_{td}} M_n + M_m$$

where k_{tc} and k_{td} represent the rate constants for termination via combination and disproportionation, respectively. However, these two rate constants are generally combined into a single rate constant k_t. The rate of termination is given by the following expression:

$$R_t = -d[M^\bullet]/dt = 2k_t[M^\bullet]^2 \quad (3.2)$$

Early on in a polymerization reaction, the rates of initiation and termination become equal, thus resulting in a steady-state concentration of free radicals, and so $R_i = R_t$. The rate equations for initiation and termination may then be combined to give the following:

$$[M^\bullet] = (fk_d[I]/k_t)^{1/2} \quad (3.3)$$

For the propagation stage, the series of (propagation) steps can be generalized in one reaction, as follows:

$$M_n^\bullet + M \longrightarrow M_{n+1}^\bullet$$

A single rate constant, k_p, is assumed to apply to these steps since the radical reactivity is effectively independent of the size of the growing polymer. Since propagation is the stage that involves the major consumption of the monomer, the rate of monomer loss can be expressed in terms of propagation only:

$$R_p = -d[M]/dt = k_p[M][M^\bullet] \quad (3.4)$$

Substitution for $[M^\bullet]$ therefore gives the following expression:

$$R_p = k_p(fk_d[I]/k_t)^{1/2}[M] \quad (3.5)$$

SAQ 3.2

A 1 M solution of styrene in benzene was polymerized by a free-radical mechanism, using an initiator concentration of 0.05 M. The steady state propagation rate was 1.5×10^{-7} mol l^{-1} s^{-1}.

(a) Calculate the propagation rate for a monomer concentration of 2 M, given that the other conditions are the same as above.

(b) Calculate the rate of propagation for an initiator concentration of 0.1 M, again under the same conditions as above.

During a free-radical polymerization, the 'reactivity' of a radical may be transferred to another species – this process is known as *chain transfer*. Such reactions involve stopping the growth of the (growing) chain radical and starting a new one in its place, as represented by the following:

$$R'H + M_x^\bullet \xrightarrow{k_{tr}} M_x + R'^\bullet$$

$$R'^\bullet + M \xrightarrow{k_a} RM^\bullet$$

where R′H is known as a *chain-transfer agent*. Mercaptans (R′SH) are examples of chain-transfer agents which lower the average polymer chain length. The transfer to the polymer or monomer with the subsequent polymerization of the double bond leads to the formation of branched molecules.

The polymerization of certain monomers, undiluted or in concentrated solution, is accompanied by a marked deviation from first-order kinetics in the direction of an increase in reaction rate and molecular weight. This deviation is known as *autoacceleration* or the *Trommsdorff–Norrish gel effect*. Such an effect results from a reduction in the rate at which the polymer molecules diffuse through the increasingly viscous mixture. This means that the ability of the polymer radicals to interact and terminate is reduced. The decrease in the termination rate leads to an increase in the polymerization rate and the molecular weight. At high conversions, the polymerization rate decreases to a low value as the mixture becomes glassy and monomer is no longer supplied to the growing polymer chain. This gel effect is most notable for acrylic polymers such as poly(methyl methacrylate) (PMMA), poly(methyl acrylate) or poly(acrylic acid). Such behaviour is illustrated for PMMA in Figure 3.1, which shows the conversion to polymer as a function of time at 20°C.

Inhibitors and retarders are substances used to control the rates of free-radical polymerizations. A *retarder* is a substance that can react with a radical to form products incapable of adding monomer. This material reduces the concentration of radicals and shortens their average lifetimes, and thus shortens the chain length. An *inhibitor* is a retarder that is particularly effective and allows no polymer to be initially formed. Examples of inhibitors and retarders include nitrobenzene, quinones and oxygen. The effects of retarders and inhibitors upon free-radical polymerization are illustrated in Figure 3.2.

3.2.2 Ionic Chain Polymerization

Ionic chain polymerization provides a viable alternative to free-radical polymerization. This type of process can take place when the reactive centre at the growing end of the polymer chain is ionic in character. For *cationic* polymerizations, the monomers contain electron-releasing substituents, such as –OR, vinyl groups and phenyl groups, while for *anionic* polymerizations the monomers contain electron-attracting substituents, such as –CN, –COOH, –COOR and halogens.

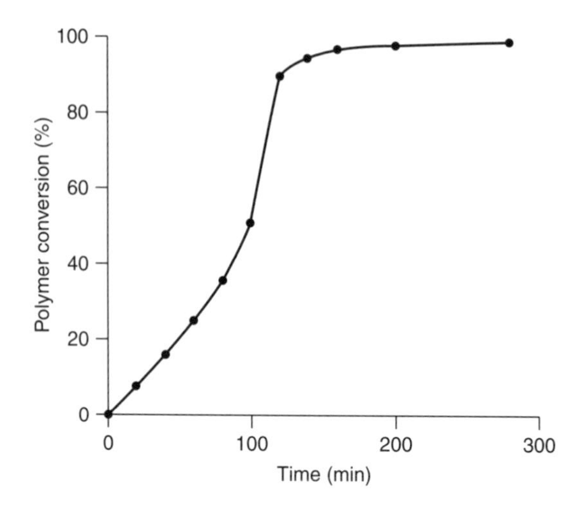

Figure 3.1 The autoacceleration effect, as displayed in the polymerization of methyl methacylate at 20°C.

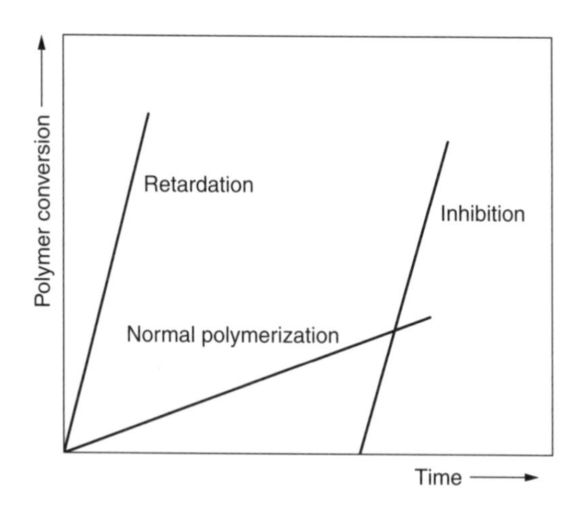

Figure 3.2 The effects of inhibitors and retarders on free-radical polmerization processes.

During the initiation stage of a cationic polymerization, cationic-active centres are created by the reaction of the monomer with an electrophile. Examples of electrophiles that may be commonly encountered include protonic acids, such as sulfuric acid:

$$CH_2{=}CHX + H_2SO_4 \longrightarrow \underset{\text{carbonium ion}}{CH_3C^+HX} + \underset{\text{counter-ion}}{HSO_4^-}$$

Lewis acids, such as boron trifluoride, aluminium chloride or tin chloride (*Friedel–Crafts catalysts*), can also be employed, although these electrophiles must be used with a co-catalyst, such as water or an organic halide. For instance:

$$BF_3 + H_2O \longleftrightarrow H^+(BF_3OH)^-$$

$$H^+(BF_3OH)^- + CH_2{=}CHX \longrightarrow CH_3C^+HX(BF_3OH)^-$$

The general reaction that takes place when using a Lewis acid is given by the following:

$$R^+A^- + CH_2{=}CHX \longrightarrow RCH_2C^+HXA^-$$

where R^+ is an electrophile and A^- is the counter-ion. The propagation steps in a cationic polymerization reaction proceed mainly via successive head-to-tail additions of the monomer to the active centre, as follows:

$$\sim CH_2C^+HXA^- + CH_2{=}CHX \longrightarrow \sim CH_2CHXCH_2C^+HXA^-$$

Termination of a cationic polymerization reaction may occur by chain transfer to a counter-ion, according to the following:

$$\sim CH_2C^+HXA^- \longrightarrow \sim CH{=}CHX + H^+A^-$$

Alternatively, termination may occur via chain transfer to the monomer, as follows:

$$\sim CH_2C^+HXA^- + CH_2{=}CHX \longrightarrow \sim CH{=}CHX + CH_3C^+HXA^-$$

SAQ 3.3

Ethylene may be polymerized by using sulfuric acid. Write down the reaction scheme for this particular process.

In anionic polymerization, a common method of initiation for the anionic chains involves the addition of a negative ion to a monomer. Initiation can involve organolithium compounds, such as ethyl lithium or butyl lithium, or bases such as potassium amide or potassium cyanide, as illustrated by the following:

$$KNH_2 \longleftrightarrow K^+ + NH_2{}^-$$

$$NH_2{}^- + CH_2{=}CHX \longrightarrow H_2NCH_2C^-HX$$

Propagation then proceeds according to the following:

$$H_2N\!\left(CH_2\text{–}CHX\right)_nCH_2C^-HX + CH_2{=}CHX$$

$$\longrightarrow H_2N\text{–}CH_2\text{–}CHX_{n+1}\text{–}CH_2C^-HX$$

The unterminated polymer can be stored at this point as a *living polymer* and then further reacted at a later stage. The propagating polymer retains active carbanionic end-groups and if more monomer is added the chains will grow further. This is feasible because the termination step involves transfer to a species not essential to the reaction at this stage. The living polymer can be terminated by quenching with water or carbon dioxide. Termination can also occur by chain transfer to the solvent:

$$H_2N{-}CH_2{-}CHX_x{-}CH_2C^{-}HX + NH_3$$
$$\longrightarrow H_2N{-}CH_2{-}CHX_x{-}CH_2CH_2X + NH_2^{-}$$

SAQ 3.4

Acrylonitrile can be polymerized in liquid ammonia by using KNH_2 as an initiator. What is the mechanism for this process? Write down the reaction scheme for this particular reaction.

3.2.3 Coordination Polymerization

Chain polymerization can occur via coordination, with organometallic complexes generally being used as catalysts for such reactions. *Ziegler–Natta catalysts* i.e. complexes formed between main-group metal alkyls and transition metal salts, are often used for these polymerizations. Table 3.1 lists the components of some typical Ziegler–Natta catalysts. The constituents of such catalysts combine together to produce a vacant coordination site on the transition metal to which a monomer molecule may bond. Another monomer molecule then attaches itself to a second vacant coordination site, from which it is able to react with the first coordinated monomer molecule. This allows the second coordination site to become vacant again and so permits another monomer molecule to enter and react with the growing polymer. The structure postulated for a metal halide–metal alkyl complex of a typical coordination catalyst is shown in Figure 3.3.

Table 3.1 Components of some typical Ziegler–Natta catalysts

Main-group metal alkyl	Transition metal salt
$(CH_3CH_2)_3Al$	$TiCl_4$
$(CH_3CH_2)_2AlCl$	VCl_3
C_4H_9Li	$MoCl_5$
C_4H_9MgI	WCl_6

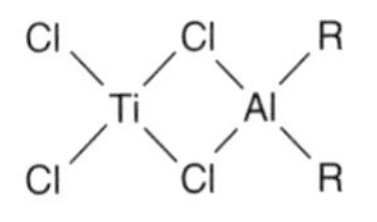

Figure 3.3 Structure of a typical coordination polymerization catalyst, where R is an alkyl group.

3.2.4 Ring-Opening Polymerization

Polymers with the general structure –R–X_n–, where –X– is a linking group such as –O–, –CO–O– or –NH–CO–, can be prepared by a ring-opening polymerization reaction. The mechanism for this type of polymerization can be represented by the following:

$$n \left(\begin{matrix} \text{R} \\ \text{Z} \end{matrix} \right) \longrightarrow \left[\text{R–Z} \right]_n$$

The driving force for the ring opening of cyclic monomers is the release of bond angle strain or steric repulsions between atoms crowded into the centre of the ring. An example of a ring-opening polymerization reaction is the formation of nylon 6 (polycaprolactam) from caprolactam, as follows:

$$n \left(\begin{matrix} (\text{CH}_2)_5 \\ \text{C(=O)–NH} \end{matrix} \right) \longrightarrow \left[(\text{CH}_2)_5\text{–NH–C(=O)} \right]_n$$

Ring opening is usually initiated by acids or bases, with the polymerization process following an ionic chain mechanism.

DQ 3.1

Predict the polymers formed from the ring-opening polymerizations of the following cyclic compounds:

(a) lactone

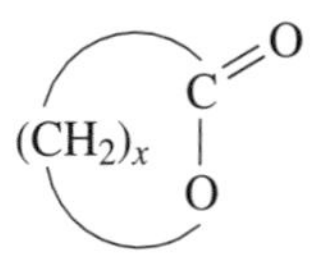

(b) cyclic ether

Answer

(a) Polyesters can be formed from lactones:

$$\left[(CH_2)_x\overset{\overset{\displaystyle O}{\|}}{C}O \right]$$

(b) Polyethers can be formed from cyclic ethers:

$$\left[(CH_2)_xO \right]$$

3.2.5 Practical Methods of Chain Polymerization

Chain reactions are employed to prepare a variety of polymers of commercial importance, with four methods being used in common practice, namely bulk, solution, suspension and emulsion polymerizations [3]. In *bulk polymerization*, the starting material consists mainly of pure monomer, plus a small amount of a monomer-soluble initiator. A schematic diagram of the reaction vessel used to carry out bulk polymerization is shown in Figure 3.4. This method can be used for the production, for example, of polyethylene, polystyrene and poly(methyl methacrylate). Bulk polymerization processes have the advantage of a high concentration of monomer, which produces high rates and degrees of polymerization. There is also the advantage of a minimum of contamination. However, there are problems with high concentrations of monomer as the viscosity of the reaction

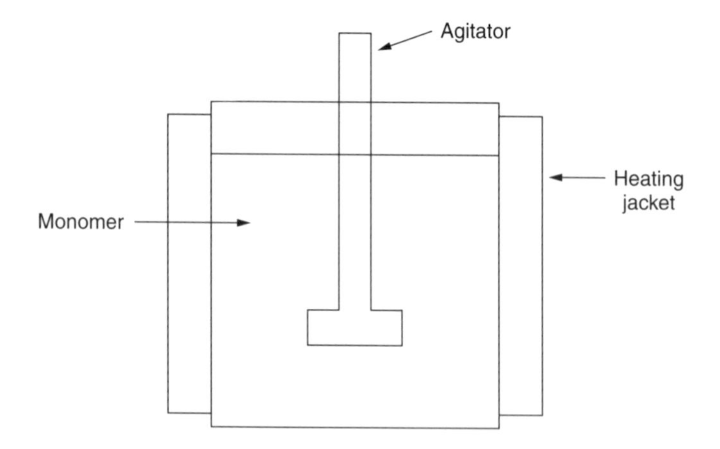

Figure 3.4 Schematic of the reaction vessel used to carry out bulk polymerization.

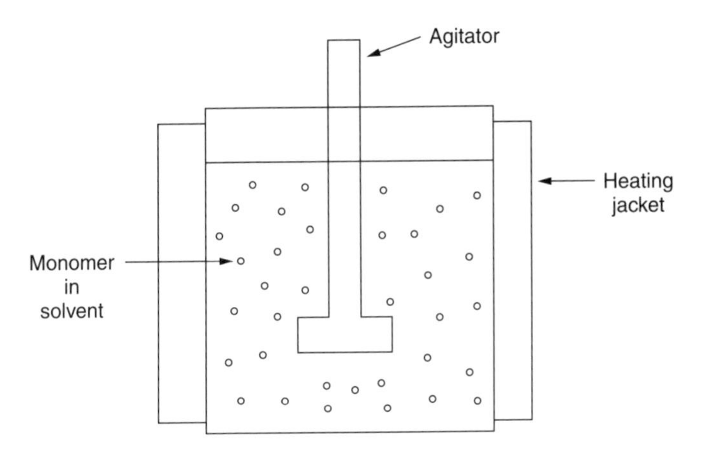

Figure 3.5 Schematic of the reaction vessel used to carry out solution polymerization.

mixture increases as the polymerization reaction proceeds. This causes difficulties in handling the product. As these polymerization reactions tend to be strongly exothermic in nature, the increasing viscosity inhibits heat dissipation and localized overheating may cause charring or degradation.

Another approach is to employ *solution polymerization*, which involves dissolving the monomer in an appropriate solvent (Figure 3.5). The presence of a solvent allows for a reduction in viscosity and heat transfer can be better controlled. In addition, this technique may produce a solution suitable for casting. However, the scope of the method is narrowed by the fact that the reaction temperature is limited by the boiling point of the solvent being employed, which in turn limits the rate of the reaction. It can be difficult to remove the last traces of the solvent, which may lead to chain transfer to the latter and consequently a restriction on the molecular weight of the final product.

Suspension polymerization occurs when the reaction mixture is suspended as droplets in an inert medium. This method can be used for the preparation of, for example, poly(vinyl chloride), polystyrene and poly(methyl methacrylate). The suspension allows for a reduction in viscosity and the heat transfer can be better controlled. However, continuous agitation is required in order to maintain the monomer in suspension and washing and drying of the sample is also necessary. The system used to carry out suspension polymerization is illustrated schematically in Figure 3.6.

Emulsion polymerization is a widely used technique and is employed in the production of a wide range of polymers including elastomers and paints. The reaction vessel used for an emulsion polymerization reaction is shown in Figure 3.7. In this method, droplets of monomer are dispersed in water with the aid of an emulsifying agent or a detergent. Typical emulsifiers used include sodium dodecyl sulfate (SDS) ($CH_3(CH_2)_{11}SO_4{}^-Na^+$) and sodium palmitate

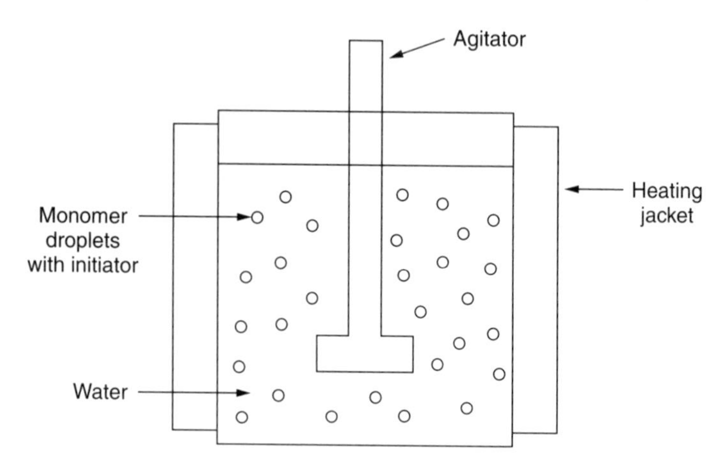

Figure 3.6 Schematic of the reaction vessel used to carry out suspension polymerization.

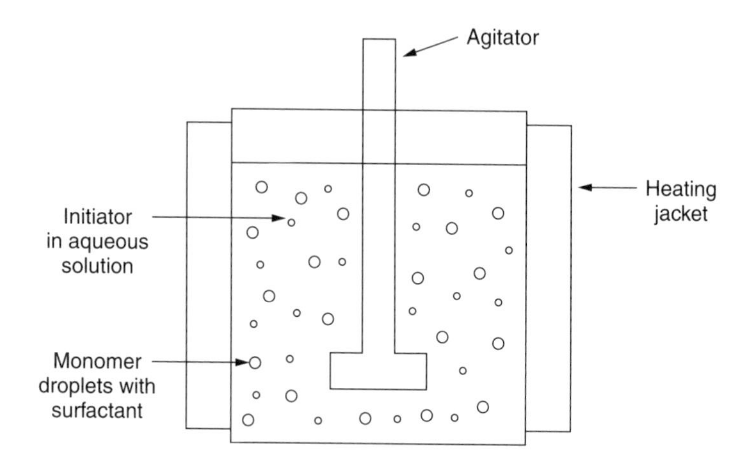

Figure 3.7 Schematic of the reaction vessel used to carry out emulsion polymerization.

($CH_3(CH_2)_{14}COO^-Na^+$). The detergent forms small micelles of the order of 10–100 μm in diameter, which contain a small quantity of monomer. Emulsion polymerization is initiated by using a water-soluble initiator such as potassium persulfate ($K_2S_2O_8$), which forms sulfate ion radicals in the water phase. These radicals enter the micelles and initiate polymerization. The monomer then diffuses from the droplets, via the water phase, into the micelles and the micelles grow into particles (Figure 3.8). Emulsion polymerization produces high-molecular-weight polymer, with the viscosity remaining relatively low. However, there are some disadvantages, including the possibility of contamination with emulsifier, and washing and drying of the product are therefore necessary.

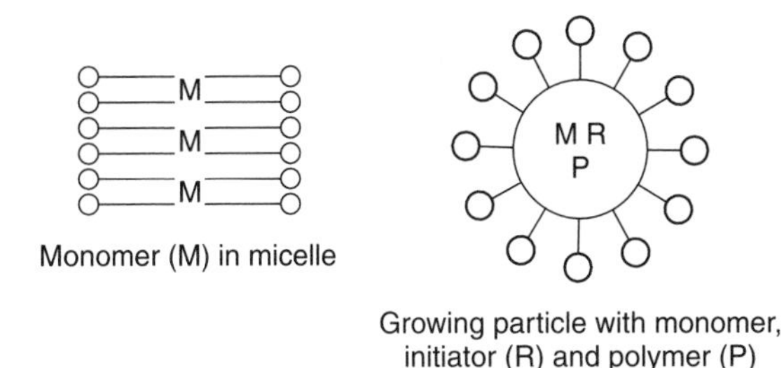

Figure 3.8 Illustration of the mechanism of emulsion polymerization.

3.3 Step Polymerization

Step polymerizations involve successive reactions between pairs of reactive functional groups on the monomers [1, 2]. The characteristic features of this type of polymerization process are as follows:

- Growth occurs throughout the matrix.
- There is a rapid loss of the monomer species.
- The molecular weight slowly increases throughout the reaction.
- The same mechanism operates throughout the reaction.
- The polymerization rate decreases as the number of functional groups decreases.
- No initiator is required to start the reaction.

Step polymerization reactions where small molecules are eliminated are termed *polycondensations*. These polymerizations involve the reaction between an organic base (such as an alcohol or an amine) and an organic acid (such as a carboxylic acid or an acid chloride) in which small molecules (such as water) are eliminated. An example of a polycondensation reaction is the formation of polyesters, as follows:

$$n\text{HOOC–R}_1\text{–COOH} + n\text{HO–R}_2\text{–OH} \longrightarrow \text{H–OOC–R}_1\text{–COO–R}_{2n}\text{–OH} + (2n - 1)\text{H}_2\text{O}$$

Step polymerization reactions where monomers react without the elimination of other molecules are termed *polyadditions*. An example of such a reaction is the formation of polyurethanes, as follows:

$$n\text{O=C=N–R}_1\text{–N=C=O} + n\text{HO–R}_2\text{–OH} \longrightarrow \text{–CO–NH–R}_1\text{–NH–CO–O–R}_2\text{–O}_n\text{–}$$

3.3.1 Statistics

Carothers developed a simple method of analysis for predicting the molecular weights of polymers prepared by step polymerization reactions. If N_0 is the original number of molecules present in an A–B monomer system and N is the number of all molecules remaining after time t, then $(N_0 - N)$ is the number of functional groups of either A or B which have reacted. At time t, the *extent of the reaction* (p) is given by the following:

$$p = N_0 - N/N_0 \quad (3.6)$$

or

$$N = N_0(1 - p) \quad (3.7)$$

Since the number-average chain length is $\overline{x}_n = N_0/N$, then:

$$\overline{x}_n = 1/(1 - p) \quad (3.8)$$

This is known as the *Carothers equation.*

The theory of Carothers is restricted to the prediction of number-average quantities. However, a simple statistical analysis first described by Flory and based upon the random nature of the step polymerization reaction allows the prediction of size distributions. The probability $P(x)$ of the existence of a molecule consisting of exactly x monomer units at time t, when the extent of reaction is p, is given by the following equation:

$$P(x) = (1 - p)p^{(x-1)} \quad (3.9)$$

The weight fraction of $x$mers (w_x) can also be related to the extent of reaction by the following:

$$w_x = x(1 - p)^2 p^{(x-1)} \quad (3.10)$$

Equations (3.9) and (3.10) are known as *Shultz–Flory distributions*. Examples of such distributions for various extents of reaction in a linear step polymerization reaction are illustrated in Figure 3.9. The weight-average chain length $(\overline{x}_{\mathrm{w}})$ can also be shown to be given by the following relationship:

$$\overline{x}_{\mathrm{w}} = (1 + p)/(1 - p) \quad (3.11)$$

3.3.2 Kinetics

If equal reactivity of the functional groups is assumed, the kinetics of step polymerization reactions are simplified, as a single rate constant can then be applied. Most step polymerizations involve bimolecular reactions that are often catalysed, with the general reaction being as follows:

$$\sim\mathrm{A} + \mathrm{B}\sim + \text{catalyst} \longrightarrow \sim\mathrm{AB}\sim + \text{ catalyst}$$

In the presence of a catalyst, the rate is accelerated, according to the following:

$$-\mathrm{d}[\mathrm{A}]/\mathrm{d}t = k[\mathrm{A}][\mathrm{B}][\text{catalyst}] \quad (3.12)$$

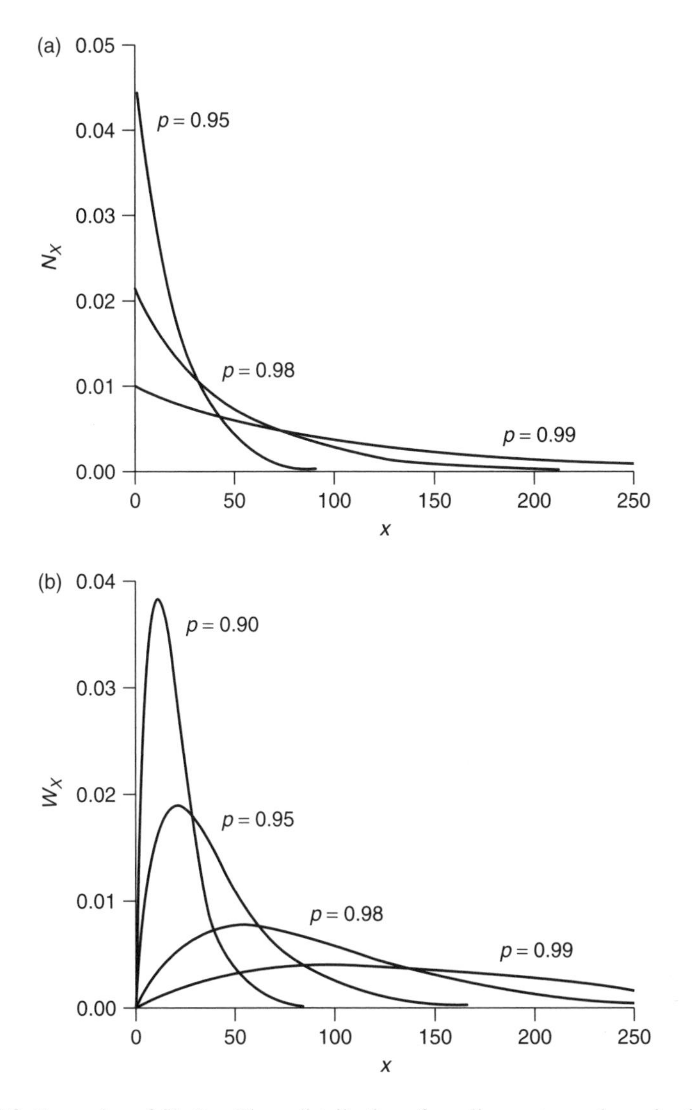

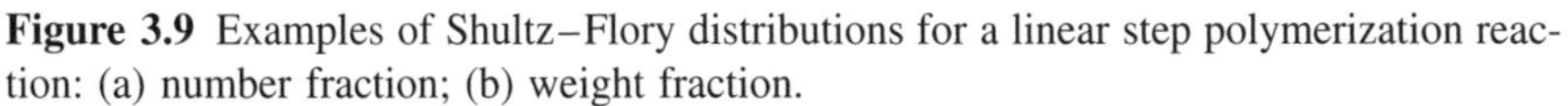

Figure 3.9 Examples of Shultz–Flory distributions for a linear step polymerization reaction: (a) number fraction; (b) weight fraction.

or

$$-\mathrm{d}[\mathrm{A}]/\mathrm{d}t = k'[\mathrm{A}][\mathrm{B}] \tag{3.13}$$

where $k' = k[\text{catalyst}]$. If $[\mathrm{A}] = [\mathrm{B}] = c$, then we can write the following:

$$-\mathrm{d}c/\mathrm{d}t = k'c^2 \tag{3.14}$$

Integration then gives:

$$1/c - 1/c_0 = k't \tag{3.15}$$

This concentration–time relationship obeys second-order kinetics and a plot of $1/c$ versus t will therefore be linear. From the Carothers equation, we have the following:

$$1/(1-p) - 1 = c_0 k't \tag{3.16}$$

since $c_0/c = N_0/N$.

SAQ 3.5

A condensation reaction of a polymer was studied and the extent of the reaction was monitored as a function of time. The results are listed below in Table 3.2. Assuming that $c_0 = 3.1 \text{ mol l}^{-1}$ and that the reaction process follows second-order kinetics, determine the rate constant for the reaction.

Table 3.2 Kinetic data obtained for a condensation reaction of a polymer (SAQ 3.5)

Time (h)	Extent of reaction
0.000	0.000
0.500	0.574
1.500	0.802
2.500	0.871

Certain step polymerizations are self-catalysed because one of the functional groups also acts as a catalyst. In such reactions, the rate equation is given by the following:

$$-\mathrm{d}[\mathrm{A}]/\mathrm{d}t = k[\mathrm{A}]^2[\mathrm{B}] \tag{3.17}$$

Assume that component 'A' acts as the catalyst and $[\mathrm{A}] = [\mathrm{B}] = c$. Integration then gives:

$$1/c^2 - 1/c_0^2 = 2kt \tag{3.18}$$

This concentration–time relationship obeys third-order kinetics and a plot of $1/c^2$ against t is linear. From the Carothers equation, we have the following:

$$c = c_0(1-p) \tag{3.19}$$

and

$$1/(1-p)^2 - 1 = 2c_0^2 kt \tag{3.20}$$

SAQ 3.6

The data listed in Table 3.3 below were obtained for a polycondensation reaction. Determine both the rate constant and the order of the reaction. Was a catalyst used for the polycondensation? In addition what would be the extent of this reaction after 24 h?

Table 3.3 Kinetic data obtained for a polycondensation reaction (SAQ 3.6)

Time (h)	Concentration ($mol\ l^{-1}$)
0.0	1.76
0.5	1.14
1.0	0.91
1.5	0.78
2.0	0.69
2.5	0.63
3.0	0.58

3.4 Copolymerization

Copolymerization involves the synthesis of polymer chains containing two or more different monomers [4]. The composition of copolymers may be determined by using the *copolymer equation*. For the copolymerization of a monomer M_1 and another monomer M_2, it is assumed that the rate of addition of monomer to the growing free-radical depends only on the nature of the end-group on the radical chain. There are four possible ways in which the monomer can add, with the reactions and their corresponding rate expressions being as follows:

Reaction	Rate
$\sim M_1^\bullet + M_1 \longrightarrow \sim M_1^\bullet$	$k_{11}[M_1^\bullet][M_1]$
$\sim M_1^\bullet + M_2 \longrightarrow \sim M_2^\bullet$	$k_{12}[M_1^\bullet][M_2]$
$\sim M_2^\bullet + M_1 \longrightarrow \sim M_1^\bullet$	$k_{21}[M_2^\bullet][M_1]$
$\sim M_2^\bullet + M_2 \longrightarrow \sim M_2^\bullet$	$k_{22}[M_2^\bullet][M_2]$

The steady-state approximation can be applied to each radical type separately, i.e. $[M_1^\bullet]$ and $[M_2^\bullet]$ must each remain constant. The rate of conversion of $M_1^\bullet$ to $M_2^\bullet$ must equal the rate of conversion of $M_2^\bullet$ to $M_1^\bullet$, as follows:

$$k_{21}[M_2^\bullet][M_1] = k_{12}[M_1^\bullet][M_2] \qquad (3.21)$$

The rates of disappearance of the two types of monomer are given as follows:

$$-\mathrm{d}[\mathrm{M}_1]/\mathrm{d}t = k_{11}[\mathrm{M}_1^\bullet][\mathrm{M}_1] + k_{21}[\mathrm{M}_2^\bullet][\mathrm{M}_1] \quad (3.22)$$

$$-\mathrm{d}[\mathrm{M}_2]/\mathrm{d}t = k_{12}[\mathrm{M}_1^\bullet][\mathrm{M}_2] + k_{22}[\mathrm{M}_2^\bullet][\mathrm{M}_2] \quad (3.23)$$

Dividing equation (3.22) by equation (3.23) gives the copolymer equation, as follows:

$$\mathrm{d}[\mathrm{M}_1]/\mathrm{d}[\mathrm{M}_2] = [\mathrm{M}_1]/[\mathrm{M}_2][(r_1[\mathrm{M}_1] + [\mathrm{M}_2])/([\mathrm{M}_1] + r_2[\mathrm{M}_2])] \quad (3.24)$$

where $r_1 = k_{11}/k_{12}$ and $r_2 = k_{22}/k_{21}$. The monomer *reactivity ratios* (r_1, r_2) represent the ratios of the rate constant for a given radical adding to its own monomer, to the rate constant for its adding the other monomer: $r_1 > 1$ means that $\mathrm{M}_1^\bullet$ prefers to add to M_1, while $r_1 < 1$ means that $\mathrm{M}_1^\bullet$ prefers to add to M_2. Note that the composition of the copolymer is independent of the overall reaction rate and the initiator concentration.

The copolymer equation can be rewritten in terms of mole fractions. Let F_1 and F_2 be the mole fractions of the monomers 1 and 2 in the polymer being formed at any instant. We therefore have the following expression:

$$F_1 = 1 - F_2 = \mathrm{d}[\mathrm{M}_1]/(\mathrm{d}[\mathrm{M}_1] + \mathrm{d}[\mathrm{M}_2]) \quad (3.25)$$

If f_1 and f_2 represent the mole fractions in the monomer feed:

$$f_1 = 1 - f_2 = [\mathrm{M}_1]/([\mathrm{M}_1] + [\mathrm{M}_2]) \quad (3.26)$$

The copolymer equation can then be rewritten as follows:

$$F_1 = (r_1 f_1^2 + f_1 f_2)/(r_1 f_1^2 + 2 f_1 f_2 + r_2 f_2^2) \quad (3.27)$$

SAQ 3.7

Styrene (110 g) is copolymerized with vinyl chloride (200 g). Given that the reactivity ratios for styrene and vinyl chloride are 17 and 0.02, respectively, calculate the initial mole fraction of styrene in the copolymer.

A copolymer system is said to be *ideal* when the two radicals show the same preference for adding one of the monomers over the other:

$$k_{11}/k_{12} = k_{21}/k_{22} \quad (3.28)$$

or:

$$r_1 = 1/r_2 \quad (3.29)$$

or:

$$r_1 r_2 = 1 \quad (3.30)$$

The end-group on the growing chain has no influence on the rate of addition. The types of units are arranged at random along the chain in this case. The copolymer equation then reduces to the following:

$$d[M_1]/d[M_2] = r_1[M_1]/[M_2] \tag{3.31}$$

The copolymer system is said to be *alternating* when each radical prefers to react exclusively with the other monomer:

$$r_1 = r_2 = 0 \tag{3.32}$$

The monomers alternate regularly along the chain, and are independent of the composition of the monomer feed. The copolymer equation then further reduces to the following:

$$d[M_1]/d[M_2] = 1 \tag{3.33}$$

Most cases lie between ideal and alternating systems, i.e. $0 < r_1r_2 < 1$.

3.5 Cross-Linking

Cross-linking involves the formation of covalent bonds between polymer chains [3]. The presence of cross-links can have a significant effect on the resulting properties of the material, as illustrated by the vulcanization of natural rubber discussed earlier in Chapter 1. Uncross-linked or lightly cross-linked polymers tend to be softer and flexible, while heavily cross-linked polymers tend to be brittle and harder. Some polymers are cross-linked by the application of heat and/or pressure, while other polymers can be cross-linked via a chemical reaction occurring at room temperature.

As discussed in Chapter 1, phenolic resins can be formed from two types of prepolymers, i.e. novolaks and resoles. A basic catalyst, e.g. hexamethylenetetraamine, is added to novolaks and heat and pressure are applied. Ammonia is generated, which provides the methylene cross-links. Resoles are cross-linked after heating.

In epoxy resins, the epoxy and hydroxy groups act as reaction sites for cross-linking. These thermosets are cured at room temperature by adding amines such as diethylenetriamine or triethylenetetraamine. The epoxy rings at the ends of the two linear molecules can be reacted with ethylenediamine to form the cross-link, according to the reaction scheme shown below. The OH groups that are formed act as reaction sites for further cross-linking reactions.

The polyester and polyether intermediates formed during the synthesis of polyurethanes (PUs) readily lend themselves to cross-linking. The polyether PUs are often derived from propene oxide. The polyesters are usually terminated by hydroxy groups that act as cross-linking sites. The prepolymers may be branched

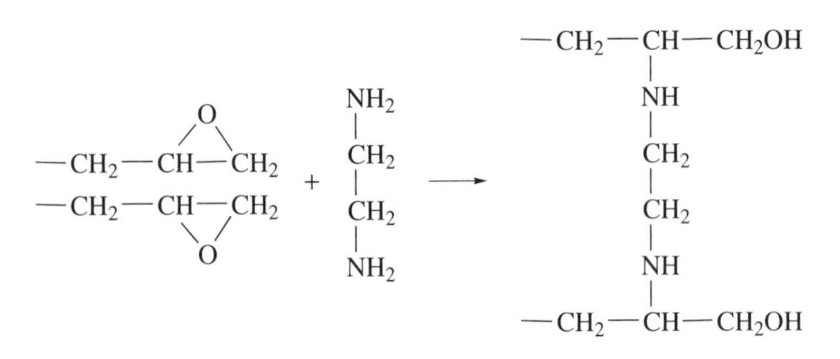

and this affects the properties of the final PU: linear polyesters form elastomers, lightly branched polyesters form flexible foams, and heavily branched polyesters form rigid foams.

The amino resins, urea–formaldehyde and melamine–formaldehyde, are formed by condensation reactions of formaldehyde with urea and melamine, respectively. The amine groups at the ends of the molecules lead to the production of highly cross-linked structures.

Unsaturated polyesters are generally cross-linked by using vinyl-type molecules, such as styrene, in the presence of a free-radical curing agent. Peroxide curing agents are commonly used and methyl ethyl ketone peroxide can be used to cross-link polyesters at room temperature. Silicone elastomers can also be cross-linked by using peroxide curing agents. For example polydimethyl siloxane can be cross-linked at room temperature by the addition of benzoyl peroxide, where the latter reacts with the two methyl groups to form Si–CH_2–CH_2–Si bridges.

3.6 Dilatometry

The technique of dilatometry can be used to monitor polymerization reactions by exploiting the fact that the density of a polymer is normally greater than that of its corresponding monomer [5]. A *dilatometer* is a vessel consisting of a capillary tube in which a liquid level can be measured with precision, as illustrated in Figure 3.10. Usually, the polymer is placed in a dilatometer containing mercury so that any volume change can be recorded as the movement of the liquid meniscus in the capillary. The volume contractions during the course of a reaction may be as large as 25%, although generally changes of the order of a few hundredths of a percent are observed. The dilatometer is placed in a water bath in order to maintain a constant temperature throughout the polymerization process.

Dilatometry can be effectively employed for the study of free-radical chain polymerizations. If low degrees of conversion are measured, the initiator concentration is assumed to be constant: in this case, a plot of $\ln[(h_0 - h_\infty)/(h_t - h_\infty)]$ versus time is linear, where h_0 is the initial height level measured in the dilatometer, h_t is the height after time t, and h_∞ is the height measured on completion of

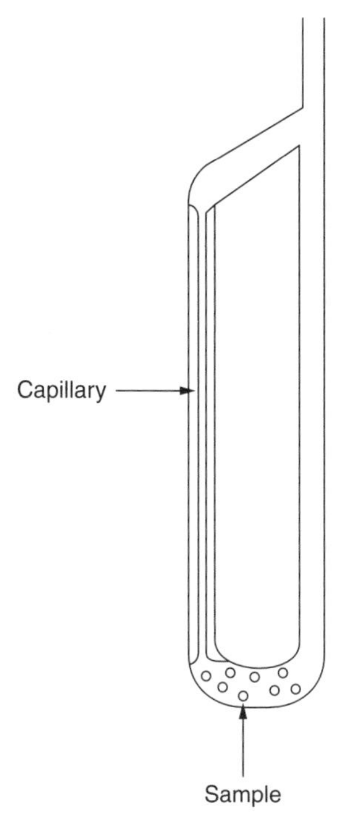

Figure 3.10 Schematic of a typical dilatometer used to monitor polymerization reactions.

the reaction. If equation (3.5) holds for the reaction (see above), then the slope of such a plot is equal to $k_p(fk_d[I]/k_t)^{1/2}$. The values of k_d, f and [I] can be determined, and so $k_p/k_t^{1/2}$ can be obtained experimentally. As such as experiment involves the study of low conversion times, the measurement of h_∞ can pose a problem. However, this parameter can be estimated by using, for instance, the Kezdy–Swinbourne method [5, 6], where the latter consists of measuring the height h at regular intervals (Δt). A plot of $h_{t+\Delta t}$ versus h_t should be linear, and a straight-line plot of $h_t = h_{t+\Delta t}$ on the same graph will intersect at $h_t = h_{t+\Delta t} = h_\infty$.

SAQ 3.8

The free-radical chain polymerization of methyl methacrylate at 80°C was monitored by using a dilatometer. The dilatometer heights measured as a function of time for this experiment are listed below in Table 3.4. A value of $h_\infty = 24.00$ mm was determined by using the Kezdy–Swinbourne method. Given that a monomer concentration of 2.00 M in toluene and a concentration of 1.35×10^{-2} M benzoyl peroxide initiator were used in the reaction, determine a value of $k_p/k_t^{1/2}$

for this polymerization. Assume that $f = 1$ and that the decomposition rate of the initiator in toluene has an approximate value of $k_d = 7 \times 10^{-5}$ s^{-1}. Compare the results you obtain with the literature values of $k_p = 1 \times 10^3$ l mol^{-1} s^{-1} and $k_t = 5 \times 10^7$ l mol^{-1} s^{-1} for the polymerization of methyl methacrylate in benzene.

Table 3.4 Dilatometry results obtained for the free-radical chain polymerization of methyl methacrylate (SAQ 3.8)

Time (s)	Height (mm)
0	0.00
890	1.85
1790	3.95
2720	5.30
3580	7.40
4500	8.25
5370	10.00

3.7 Infrared Spectroscopy

As the infrared spectra of monomers are different to those of the corresponding polymers, it is possible to use infrared spectroscopy to monitor polymerization reactions. The technique has also been applied to the study of the curing of epoxy resins [7, 8]. For example, the extent of cross-linking of epoxy resins with amines can be examined by using the C–O and C–H stretching bands because the cross-linking process involves the opening of the epoxy ring. For instance, the absorbances of the 912 and 3226 cm^{-1} bands may be measured as a function of time to follow the cross-linking reaction.

Infrared spectroscopy has been successfully applied to studying the formation of polyurethanes. Attenuated total reflectance (ATR) spectroscopy is particularly useful for monitoring the infrared spectra of reaction mixtures over short time intervals and for characterizing reaction progress [9]. For instance, the rate of consumption of free isocyanate groups and the competitive formation of urethane, isocyanurate and urea linkages can be deduced from the decrease of the N=C=O stretching mode at 2275 cm^{-1}, and the increase of the C=O stretching modes at 1725, 1710 and 1640 cm^{-1}, respectively.

3.8 Raman Spectroscopy

Polymerization kinetic studies can be carried out by using Raman spectroscopy [7, 10, 11]. Raman spectroscopy has the advantage over, say, infrared spectroscopy

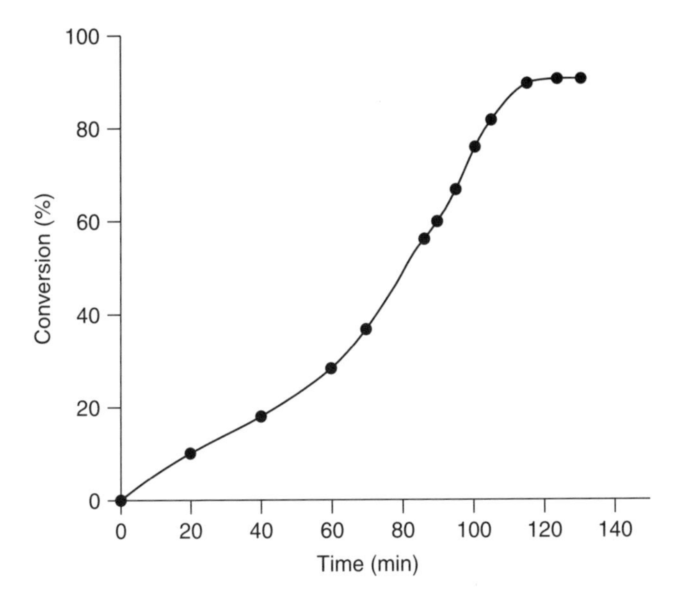

Figure 3.11 Plot of the percentage conversion versus time for the polymerization of methyl methacrylate at 65°C, using data obtained from Raman spectroscopic monitoring of the reaction.

in that glass may be used in the cell design and bulk samples can be more readily investigated. Most addition polymerizations involve the loss of C=C bonds. As C=C stretching produces a strong peak in the Raman spectrum, Raman spectroscopy is an obvious choice of technique for the quantification of such polymerization reactions. The results of a Raman study of the polymerization of methyl methacrylate at 65°C with AIBN as the initiator have been reported [12]. The plot of conversion as a function of time shown in Figure 3.11 illustrates clearly an induction period where polymerization proceeds slowly. After this period, a faster reaction then takes place. The plot shown in this figure also displays the phenomenon of autoacceleration (see Section 3.2.1 above).

SAQ 3.9

The emulsion polymerization of vinyl acetate can be studied by monitoring the decrease in the intensity of the C=C stretching mode near 1650 cm^{-1} in the Raman spectrum [13]. Figure 3.12 below shows the intensity of the C=C mode of vinyl acetate as a function of time during such an emulsion polymerization reaction. Use these data to illustrate the conversion of vinyl acetate to polymer in this reaction. The peak height can be used as a measure of conversion.

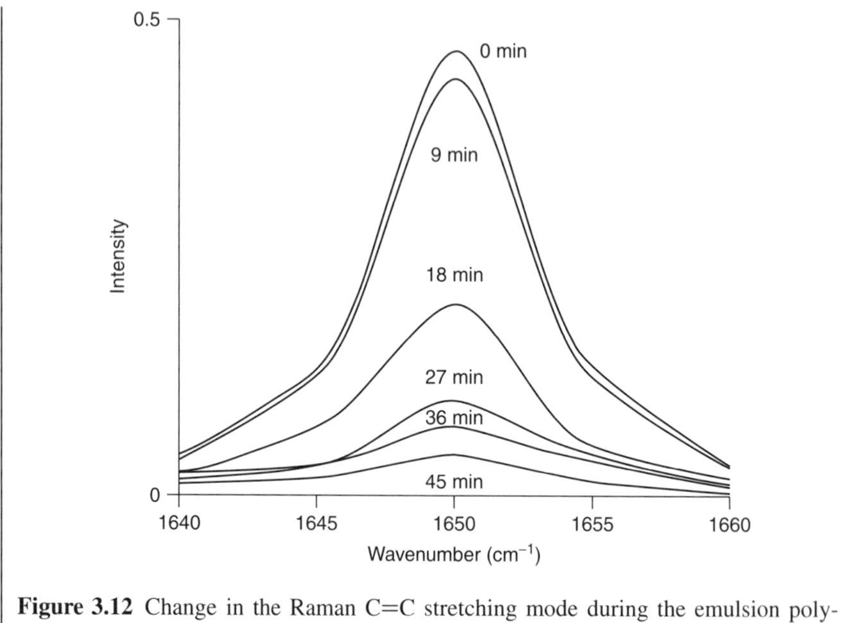

Figure 3.12 Change in the Raman C=C stretching mode during the emulsion polymerization of vinyl acetate (SAQ 3.9).

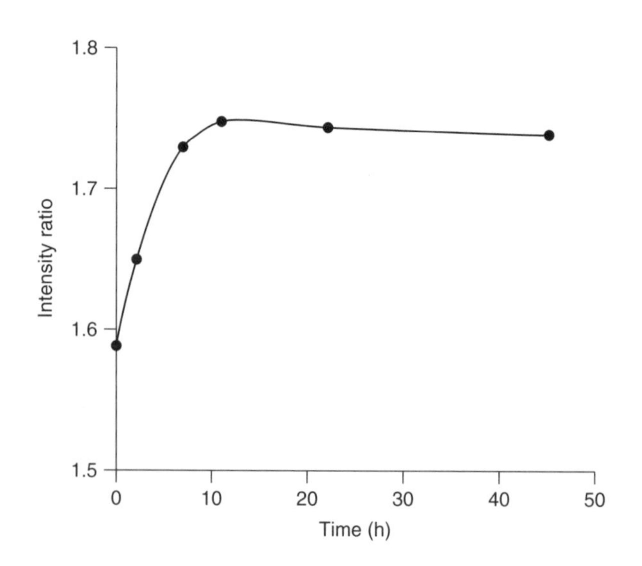

Figure 3.13 Monitoring of the curing reaction of an epoxy resin (Araldite) by using Raman spectroscopy.

Raman spectroscopy can also be used to monitor curing processes. The curing of the epoxy resin, 'Araldite', can be followed by using the intensity of the 1257 cm^{-1} mode due to C–O–C stretching [10]. This mode decreases as the reaction proceeds and the intensity can be ratioed against the aromatic ring mode at 1608 cm^{-1}, the intensity of which remains constant during curing. Figure 3.13 shows the change in the 1608 cm^{-1}/1257 cm^{-1} intensity ratio for the epoxy resin as a function of time at room temperature. This plot illustrates that the curing is substantially complete after about 10 h, and is essentially linear in nature up until that time.

3.9 Nuclear Magnetic Resonance Spectroscopy

^{13}C NMR spectroscopy has been widely used to investigate the curing of epoxy resins [7, 8]. This technique is particularly useful because the opening of the epoxide ring during the cross-linking process results in a shift downfield of the two epoxide resonances. Figure 3.14 illustrates the magic-angle spinning ^{13}C NMR spectra of an epoxy resin based on the diglycidyl ether of bisphenol A (DGEBA) during the curing process with piperidine (PIP) [14]. A comparison of the spectra of the amorphous prepolymer and the cured resin shows that, although very similar in the aromatic region, in the aliphatic region there are distinct changes on polymerization. The opening of the epoxide ring during polymerization results in downfield chemical shifts for the resonance lines 'f' and 'g'. An estimate of the degree of polymerization can be made by examination of the disappearance of these two peaks.

3.10 Differential Scanning Calorimetry

DSC is widely used to study polymerizing systems and has been particularly effective for the study of epoxy curing [15]. Such reactions are generally exothermic and so readily lend themselves to this technique. The kinetics of polymerization may be monitored by using either isothermal or scanning modes of operation. DSC traces demonstrating the curing of an epoxy resin with an amine curing agent are shown in Figure 3.15. The upper trace illustrates the first run of an uncured sample of an epoxy and shows a large exotherm which reaches a maximum at 170°C. This supports the observation that, although this reaction occurs even at low temperatures, the reaction is rapid above a temperature of 100°C.

3.11 Electron Spin Resonance Spectroscopy

Electron spin resonance (ESR) spectroscopy, also known as *electron paramagnetic resonance* (EPR) spectroscopy, is based on similar principles to NMR

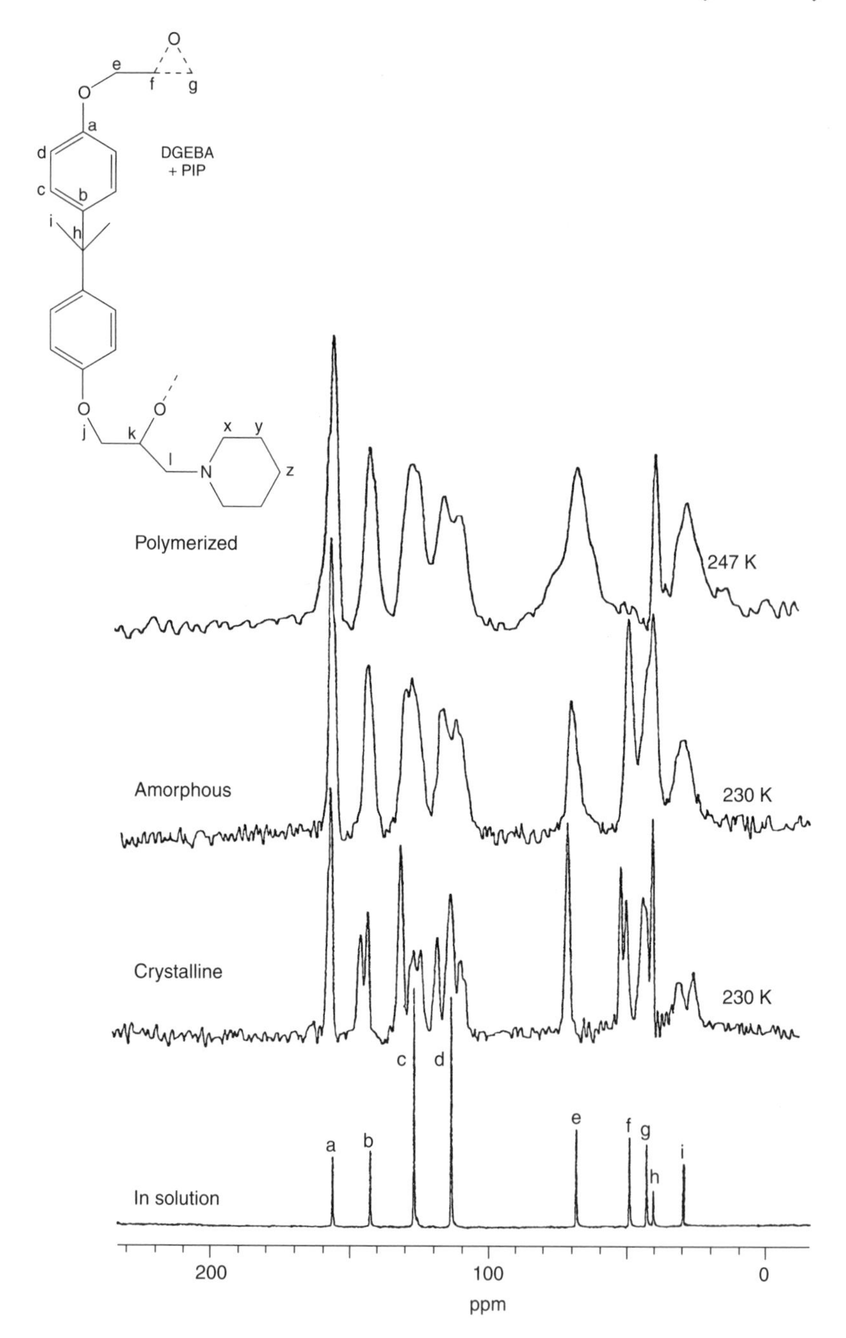

Figure 3.14 Magic-angle spinning ^{13}C NMR spectra obtained during the curing process of a DGEBA epoxy resin with piperidine. Reprinted with permission from Garroway, A. N., Ritchey, W. M. and Moniz, W. B., *Macromolecules*, **15**, 1051–1059 (1982). Copyright (1982) American Chemical Society.

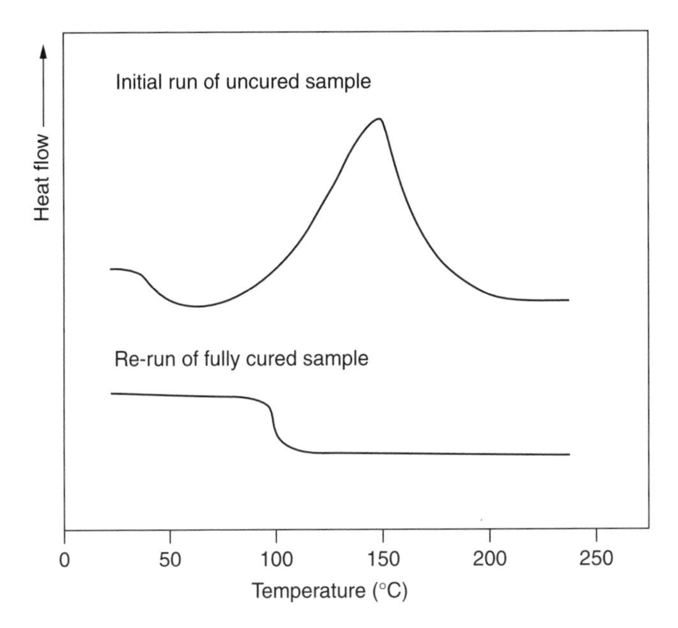

Figure 3.15 DSC traces illustrating the curing of an epoxy resin with an amine curing agent.

spectroscopy [16, 17]. However, ESR spectroscopy employs microwave rather than radiowave frequencies, and the spin transitions of unpaired electrons rather than nuclei are observed. In addition, information is obtained by using the first derivative of the absorption curve. As unpaired electrons are needed to be present in a polymer in order to be observed in an ESR spectrum, this technique is mainly used for the study of polymerization and degradation processes.

Parallelling NMR spectroscopy, in ESR spectroscopy the electron in a parallel orientation in the externally applied field with a spin quantum number of 1/2 will have only two orientations, i.e. aligned with the field or opposed to the field. The electron aligned with the field can absorb energy and change to an orientation opposed to the field (*Zeeman splitting* for an electron) when the frequency is the same as the microwave frequency (*electron spin resonance*).

The position of an ESR line is usually determined as the point when the *derivative spectrum* crosses zero. The position is usually characterized by a *g-value*, i.e. the constant of proportionality between the frequency and the field at which resonance occurs, given by the following expression:

$$\Delta E = h\nu = g\beta H_0 \tag{3.34}$$

where ΔE is the energy which will induce transitions, h is the Planck constant, ν is the frequency, g is the g-value (=2.002 319 for unbound electrons), β is the Bohr magneton and H_0 is the strength of the applied external magnetic field.

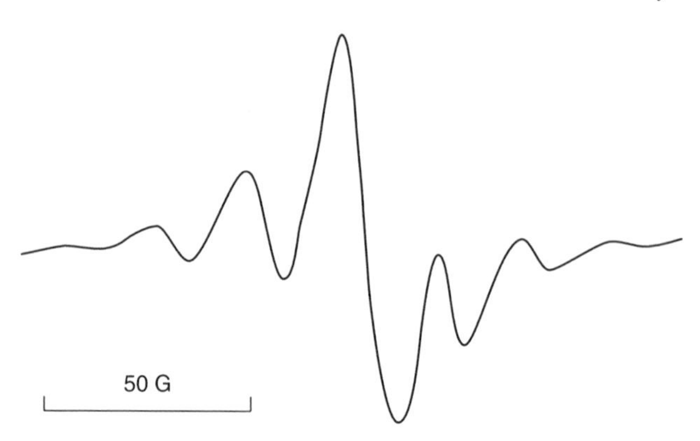

Figure 3.16 ESR spectrum of the radical species formed during the ultraviolet polymerization of methyl methacrylate with benzoyl peroxide. From Ranby, B. and Rabek, J. F., *ESR Spectroscopy in Polymer Research*, Figure 4-63, p. 104. © Springer-Verlag, 1977. Reproduced by permission of Springer-Verlag GmbH & Co. KG.

Generally, the g-value of bound unpaired electrons is not the same as that of unbound electrons.

Splitting in the energy levels of an electron arises from an increase in the number of such levels resulting from exposure of a polymer with unpaired electrons to a magnetic field. The interaction of an electron and the magnetic nucleus is called a *hyperfine interaction*, while splitting in the energy levels is known as *hyperfine splitting*. The hyperfine splitting constant is the separation between two hyperfine lines of the spectrum.

ESR spectroscopy can be used to examine a wide range of polymerization processes, including free-radical polymerization, ionic polymerization and copolymerization, as well as radiation-initiated processes. Figure 3.16 illustrates the ESR spectrum of the radical species formed by the ultraviolet irradiation of methyl methacrylate with benzoyl peroxide [16]. The spectrum is a superposition of a septet, an asymmetric singlet and a doublet from the hydrogen atoms. The asymmetric singlet and the septet are due to the benzoyl peroxy radical and a methyl methacrylate radical, respectively. When the system is heated to higher temperatures, the ESR spectrum changes. In the resulting spectrum, a septet and a quintet are observed, with the latter appearing as a result of the addition of a benzoyl peroxy radical to a methyl methacrylate monomer unit.

3.12 Refractometry

A straightforward method for monitoring polymerization processes is the measurement of refractive indices [17]. The refractive index (n) of a material is the ratio of the velocity of light in a vacuum to its velocity in the material, as defined

Table 3.5 Refractive indices of some common polymers

Polymer	Refractive index
PE	1.51–1.52
PP	1.49
PVC	1.54–1.56
PS	1.59–1.60
PMMA	1.49
Nylon	1.54
PTFE	1.35–1.38
PCTFE	1.39–1.43
PVDC	1.60–1.63
Cellulose acetate	1.46–1.50
PVA	1.47–1.49
PVAl	1.49–1.53
Phenolic resin	1.50–1.70
Natural rubber	1.52

by Snell's Law, as follows:

$$n = \sin\theta / \sin\phi \tag{3.35}$$

where θ is the angle of incidence and ϕ the angle of refraction. The refractive indices of some common polymers are listed in Table 3.5.

A refractometer can be used to measure the time-dependence of the conversion from monomer to polymer given that the refractive index of the polymer is greater than that of its corresponding monomer. Refractometry has been used to investigate the free-radical chain polymerization of methyl methacrylate via a bulk polymerization process, with some typical results being shown in Figure 3.17. The extent of the reaction (p) was calculated by using the following expression:

$$p = 1.0684(n_{\mathrm{p}} - n_1)/(0.2137n_{\mathrm{p}} - 0.2359) \tag{3.36}$$

where n_1 is the refractive index of the monomer and n_{p} is the refractive index of the reaction mixture. The shape of the curve indicates a deviation from first-order kinetics and so again illustrates the process of autoacceleration.

DQ 3.2

Explain why the curve shown in Figure 3.17 is not linear.

Answer

As the polymerization reaction proceeds, there is an increase in the viscosity of the reaction mixture which reduces the mobility of the reacting

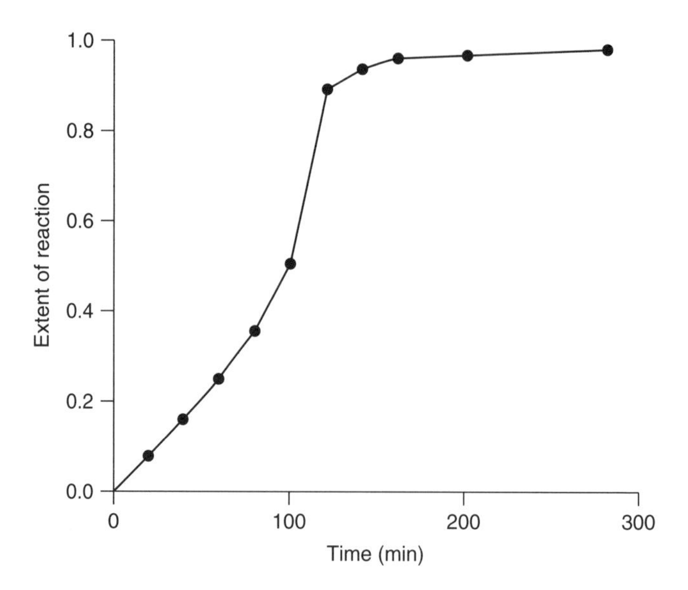

Figure 3.17 Free-radical chain polymerization of methyl methacrylate as studied by refractometry.

> *species. The growing polymer molecules are more affected by this than the monomer molecules or the fragments arising from the decomposition of the initiator. Hence, termination reactions will slow down and eventually stop, while initiation and propagation reactions will still continue. Such a decrease in the rate of the termination step leads to the observed increase in the overall rate of polymerization.*

Summary

In this chapter, the characteristics of various chain polymerization processes, including free-radical, ionic, coordination and ring-opening polymerizations, as well as step-polymerization reactions, were introduced. Certain aspects of copolymerization and cross-linking were also covered. Such polymerization reactions may be monitored by using a number of analytical techniques, including dilatometry, infrared spectroscopy, Raman spectroscopy, NMR spectroscopy, differential scanning calorimetry, ESR spectroscopy and refractometry.

References

1. Stevens, M. P., *Polymer Chemistry: An Introduction*, Oxford University Press, Oxford, UK, 1999.
2. Young, R. J. and Lovell, P. A., *Introduction to Polymers*, 2nd Edn, Chapman and Hall, London, 1991.

3. Walton, D. J. and Lorimer, J. P., *Polymers*, Oxford University Press, Oxford, UK, 2000.
4. Billmeyer, F. W., *Textbook of Polymer Science*, 3rd Edn, Wiley, New York, 1984.
5. Martin, O., Mendicuti, F. and Tarazona, M. P., *J. Chem. Educ.*, **75**, 1479–1481 (1999).
6. Connors, K. A., *Chemical Kinetics: The Study of Reaction Rates in Solution*, VCH, New York, 1990.
7. Koenig, J. L., *Spectroscopy of Polymers*, 2nd Edn, Elsevier, Amsterdam, The Netherlands, 1999.
8. Sandler, S. R., Karo, W., Bonesteel, J. and Pearce, E. M., *Polymer Synthesis and Characterization: A Laboratory Manual*, Academic Press, San Diego, CA, 1998.
9. Siesler, H. W. and Holland-Moritz, K., *Infrared and Raman Spectroscopy of Polymers*, Marcel Dekker, New York, 1980.
10. Agbenyega, J. K., Ellis, G., Hendra, P. J., Maddams, W. F., Passingham, C., Willis, H. A. and Chalmers, J., *Spectrochim. Acta, A*, **46**, 197–216 (1990).
11. Maddams, W. F., *Spectrochim. Acta, A*, **50**, 1967–1986 (1994).
12. Williams, K. P. J. and Maddams, S. M., *Spectrochim. Acta, A*, **46**, 187–196 (1990).
13. Brookes, A., Dyke, J. M., Hendra, P. J. and Strawn, A., *Spectrochim. Acta, A*, **53**, 2303–2311 (1997).
14. Garroway, A. N., Ritchey, W. M. and Moniz, W. B., *Macromolecules*, **15**, 1051–1059 (1982).
15. Haines, P. J., *Thermal Methods of Analysis: Principles, Applications and Problems*, Blackie, London, 1995.
16. Ranby, B. and Rabek, J. F., *ESR Spectroscopy in Polymer Research*, Springer-Verlag, Berlin, 1977.
17. Rabek, J. F., *Experimental Methods in Polymer Chemistry*, Wiley, Chichester, 1980.

Chapter 4

Molecular Weight

Learning Objectives

- To understand the different measures of molecular weights of polymers, and how to calculate them.
- To determine the molecular weights of polymers using data obtained from viscometry, size-exclusion chromatography, ultracentrifugation, osmometry, light scattering techniques, end-group analysis and turbidimetric titrations.

4.1 Introduction

As a result of the methods by which they are made, polymers contain mixtures of different molecular sizes. The properties of polymers are very dependent upon the average size of the molecules that are present. A largely low-molecular-weight polymer will behave in quite a different manner to a high-molecular-weight polymer. For instance, a low-molecular-weight polystyrene (PS) sample will melt at a lower temperature than a high-molecular-weight sample of the same polymer. There are a number of approaches to calculating the average molecular weights of polymers and these concepts are introduced in this present chapter. In addition, there are a variety of methods by which the molecular weights of polymers may be determined and a number of these are discussed in the following sections. Viscometry, size-exclusion chromatography, ultracentrifugation, osmometry, light scattering, end-group analysis and turbidimetric titration may all be employed to determine the molecular weights of polymer samples.

4.2 Molecular Weight Calculations

Synthetic polymers consist of a mixture of molecules with various chain lengths and molecular weights. Such mixtures are described as being *polydisperse*. A typical distribution of polymer molecular weights is shown in Figure 4.1. The exact distribution of the molecular weights will depend on the method of preparation. This figure also illustrates two common ways of defining the polymer molecular weight. First, there is the *number-average molecular weight* ($\overline{M}_n$). This is determined by dividing the chains into a series of size ranges and then determining the number fraction of chains within each range by using the following equation:

$$\overline{M}_n = \Sigma x_i M_i \tag{4.1}$$

where M_i is the mean molecular weight of the size range i and x_i is the fraction of the total number of chains within this range.

SAQ 4.1

The molecular weight distribution of a PVC sample is shown below in Table 4.1. Calculate the number-average molecular weight of this material.

Table 4.1 Molecular weight distribution of a PVC sample (SAQ 4.1)

Mean M_i (g mol^{-1})	x_i
7500	0.05
12 500	0.16
17 500	0.22
22 500	0.27
27 500	0.20
32 500	0.08
37 500	0.02

Secondly, there is the *weight-average molecular weight* ($\overline{M}_w$), which is defined as the weight fraction of molecules within the various size ranges, as given by the following equation:

$$\overline{M}_w = \Sigma w_i M_i \tag{4.2}$$

where M_i is the mean molecular weight in the size range i and w_i is the weight fraction of the molecules within this range.

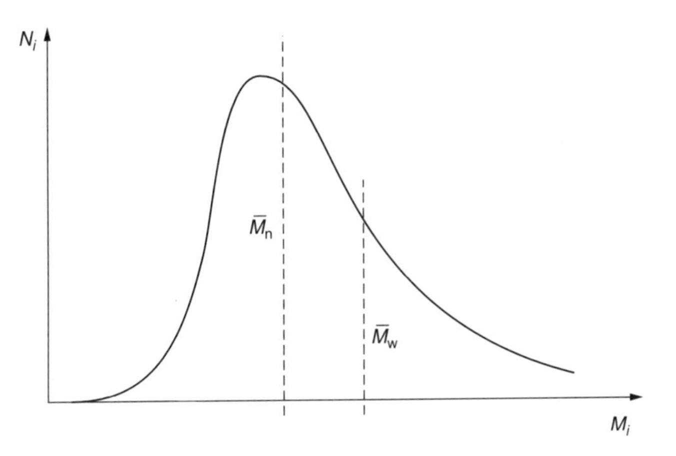

Figure 4.1 A typical distribution of polymer molecular weights.

SAQ 4.2

The data shown below in Table 4.2 were obtained for the molecular weight distribution of a sample of poly(vinyl alcohol). Determine the weight-average molecular weight of this material.

Table 4.2 Molecular weight distribution of a sample of poly(vinyl alcohol) (SAQ 4.2)

Mean M_i (g mol^{-1})	w_i
7500	0.02
12 500	0.10
17 500	0.18
22 500	0.29
27 500	0.26
32 500	0.13
37 500	0.03

The number-average degree of polymerization ($\overline{n}_n$) is given by the following:

$$\overline{n}_n = \overline{M}_n/m \tag{4.3}$$

where m is the molecular weight of the structural repeat unit. Correspondingly, the weight-average degree of polymerization is given by:

$$\overline{n}_w = \overline{M}_w/m \tag{4.4}$$

For a copolymer, m is determined from the following:

$$m = \Sigma f_j m_j \tag{4.5}$$

where f_j and m_j are, respectively, the chain fraction and molecular weight of the repeat unit j.

SAQ 4.3

The number-average molecular weight of a polystyrene sample is 500 000 g mol^{-1}. Calculate the number-average degree of polymerization of this material.

Monodisperse polymers contain molecules which are all of the same molecular weight, and $\overline{M}_n = \overline{M}_w$. In all other cases, $\overline{M}_w > \overline{M}_n$. The *polydispersity index* (PDI) is a measure of the breadth of the molecular weight distribution and is given by the following expression:

$$\text{PDI} = \overline{M}_w / \overline{M}_n \tag{4.6}$$

For step polymerizations, the molecular weights can be obtained by using the following equations:

$$\overline{M}_n = M_0/(1 - p) \tag{4.7}$$

$$\overline{M}_w = M_0(1 + p)/(1 - p) \tag{4.8}$$

where M_0 is the molecular weight of the repeat unit of the polymer and p is the extent of reaction. An expression for the PDI can also be obtained in terms of conversion, as follows:

$$\overline{M}_w / \overline{M}_n = 1 + p \tag{4.9}$$

SAQ 4.4

For a polyamide with the structural repeat unit–$[NH(CH_2)_6NHCO(CH_2)_4CO]_n$–, the percentage conversion of the functional groups was found to be 96%. Calculate the number- and weight-average molecular weights of this polymer.

4.3 Viscometry

For the determination of molecular weights of polymers by viscosity methods, dilute solutions (~1%) are used [1, 2]. Two important parameters are utilized in viscosity techniques. The first of these, the *relative viscosity*, is the ratio of the viscosity of the solution (η) to the viscosity of the pure solvent (η_0), i. e. η/η_0. Secondly, the *reduced viscosity* is defined as $\eta - \eta_0/\eta_0 c$, where η is the viscosity of a solution of concentration c (given in g l^{-1}). The relative viscosity varies with concentration as a power series, as follows:

$$\eta/\eta_0 = 1 + \eta c + k\eta^2 c^2 + \cdots, \text{etc.} \tag{4.10}$$

For low concentrations ($\leq 1\%$), the terms above c^2 can be ignored and the *Huggins equation* is obtained:

$$(\eta - \eta_0)/\eta_0 c = \eta + k\eta^2 c \tag{4.11}$$

The intrinsic viscosity, $[\eta]$, is determined by taking the following limit:

$$[\eta] = \lim_{c \to 0}[(\eta - \eta_0)/\eta_0 c] \tag{4.12}$$

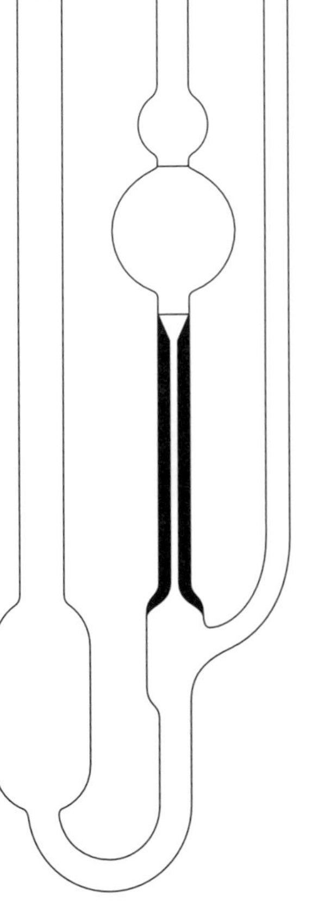

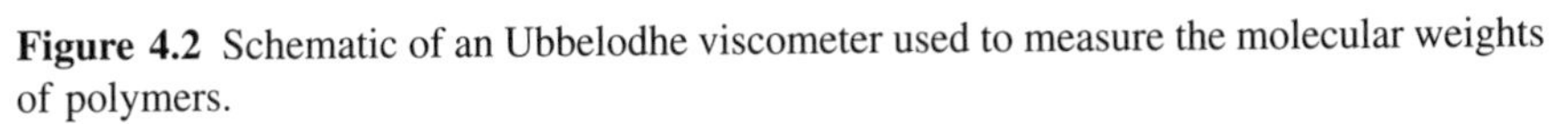

Figure 4.2 Schematic of an Ubbelodhe viscometer used to measure the molecular weights of polymers.

The value of $[\eta]$ may be experimentally determined by plotting $((\eta - \eta_0)/\eta_0 c)$ as a function of c, or by plotting $(\eta/\eta_0 - 1)/c$ versus c and finding the y-intercept.

Viscosity may be measured by using a *viscometer*, with the most commonly employed being the Ubbelodhe viscometer, as illustrated in Figure 4.2. The time taken for the solution to flow through the capillary is measured and then compared with a standard sample. The ratio of the efflux time of the solution (t) to that of the pure solvent (t_0) may be taken as being equivalent to the ratio of the viscosities obtained from the Huggins equation, as follows:

$$(t - t_0)/t_0 c = (\eta - \eta_0)/\eta_0 c \tag{4.13}$$

The relative molecular weight may be obtained by using the *Mark–Houwink–Sakurada equation* (also known as the *Mark–Houwink equation*), given as follows:

$$[\eta] = KM^a \tag{4.14}$$

where K and a are constants which are dependent on the solvent, the type of polymer and the temperature of the system.

SAQ 4.5

The viscosities of a series of solutions of polystyrene in toluene were measured at 25°C and the results obtained are summarized below in Table 4.3. Determine the molecular weight of this polymer. The Mark–Houwink–Sakurada constants for this system are $K = 3.80 \times 10^{-5}$ l g^{-1} and $a = 0.63$.

Table 4.3 Viscosity measurements obtained for a series of polystyrene solutions in toluene (SAQ 4.5)

Concentration (g l^{-1})	0	2.0	4.0	6.0	8.0	10.0
Viscosity (10^{-4} kg m^{-1}s^{-1})	5.58	6.15	6.74	7.35	7.98	8.64

4.4 Chromatography

Gel permeation chromatography (GPC), a type of *size-exclusion chromatography* (SEC), is a technique that employs porous non-ionic gel beads to separate polymers in solution [2–4]. Beads containing pores of various sizes and distributions are packed into a column in SEC. Such beads are commonly made of glass or cross-linked PS. A solvent is pumped through the column and then a polymer solution in the same solvent is injected into the column. Fractionation of the polymer sample results as different-sized molecules are eluted at different times. A schematic diagram of a typical size-exclusion chromatograph is shown in Figure 4.3.

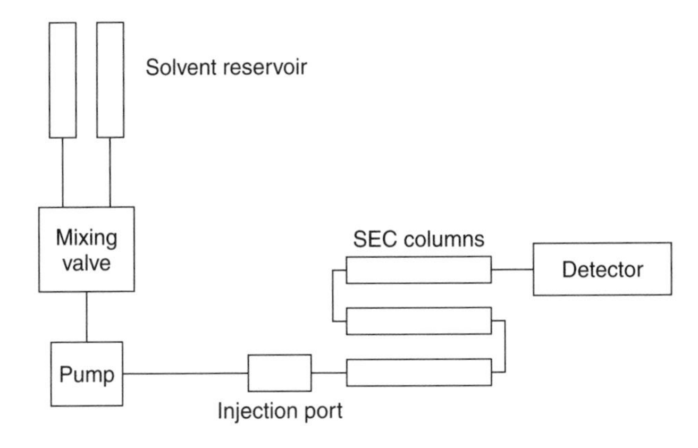

Figure 4.3 Schematic of a typical size-exclusion chromatograph system.

Fractionation of molecules in SEC is governed by hydrodynamic volume rather than by molecular weight. The largest polymers in the solution cannot penetrate the pores within the cross-linked gel beads, and so they will elute first as they are excluded and their retention volume is smaller. The smallest polymer molecules in the solution are retained in the interstices (or the voids) within the beads, and so require more time to elute as their retention volume is larger (Figure 4.4).

A size-exclusion chromatogram is a plot of detector response as a function of the retention volume (V_R). It is usual here to measure the heights (h_i) above the baseline and counts, respectively. The chromatogram is normalized before its shape is compared with a standard (reference) chromatogram as it is difficult to inject exactly the same quantity of sample into the chromatograph in each experiment.

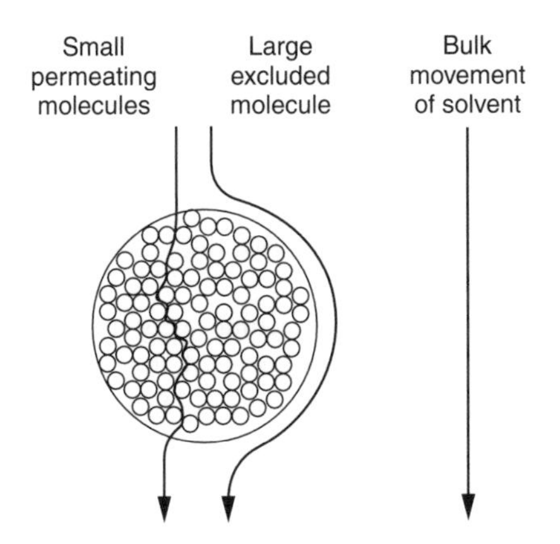

Figure 4.4 Illustration of the separation of polymer molecules of different sizes in size-exclusion chromatography.

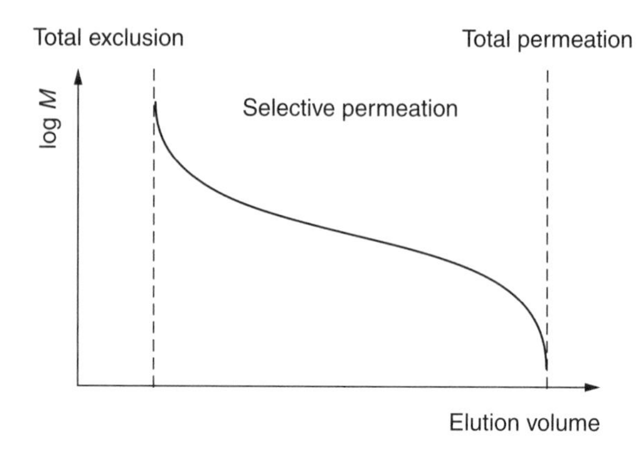

Figure 4.5 The general form of a calibration curve used in size-exclusion chromatography.

In order to obtain a molecular weight distribution, the column must be calibrated by using fractions of a known molecular weight so as to relate molecular weight to the eluted volume. Commercially available PS samples with narrow molecular weight distributions are often used as calibration standards. A calibration curve is produced by plotting the logarithm of molecular weight versus the elution volume, as illustrated in Figure 4.5. The SEC curve can be broken down into increments, where the weight fraction (w_i) of any increment is given by the following:

$$w_i = h_i/\Sigma h_i \tag{4.15}$$

The molecular weight of the i-th species (M_i) is obtained from the calibration curve at point V_i. The molecular weight averages are then calculated by using equation (4.2) and the following equation:

$$\overline{M}_{\rm n} = 1/\Sigma(w_i/M_i) \tag{4.16}$$

This approach is valid for calculating the molecular weights of monodisperse standards if the same linear polymers are available. However, such standards are often not obtainable and average molecular weights based upon PS standards are therefore commonly reported. Polymers of the same molecular weight, but of a different architecture, will elute from an SEC column at different times. For instance, linear PS will be eluted at a different time to a star-shaped PS. An alternative approach is to use a *universal calibration method.* In this, the values of $[\eta]M$ as a direct measure of the hydrodynamic volume of the polymer, (where M is the molecular weight) are taken. There is an approximately linear relationship between log $[\eta]M$ and $V_{\rm R}$ and such a universal calibration curve is

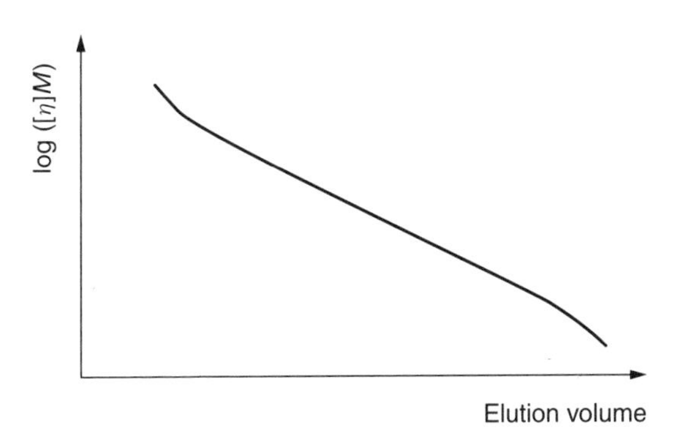

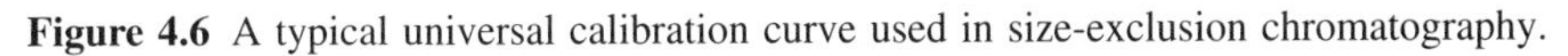
Figure 4.6 A typical universal calibration curve used in size-exclusion chromatography.

illustrated in Figure 4.6. Thus, where the Mark–Houwink–Sakurada constants are known, by using equation (4.14), V_h can be directly related to M.

DQ 4.1

A mixture of three polymer samples of similar molecular weights and chemical structures, but with different physical structures, are injected together into a size-exclusion chromatograph. One polymer is linear, another is branched and the third has a star-shaped structure. In what order will the polymers be eluted?

Answer

Chain branching will affect the hydrodynamic volume. The linear polymer is smaller and can permeate into the pores of the beads more readily that the larger branched and star molecules. The branched molecule is 'intermediate' in behaviour and can permeate some, but not all, of the passages in the column. The star-shaped polymer is a larger molecule which will not permeate the pores and thus will be excluded. Hence, the largest star-shaped polymer will be eluted first, followed by the branched polymer, and then finally the smallest linear polymer.

SAQ 4.6

A size-exclusion chromatogram was obtained for a sample of poly(methyl methacrylate) (PMMA) in tetrahydrofuran at 25°C, with the data derived from the experiment being listed below in Table 4.4. Under identical conditions, samples of PS with narrow molecular weight distributions were also examined by using SEC. A linear calibration curve was produced, with $M = 98\,000$ eluting at a retention volume of 130 ml and $M = 1800$ eluting at 165 ml. Calculate $\overline{M}_n$ and $\overline{M}_w$ for the PMMA sample.

Table 4.4 SEC data obtained for a sample of poly(methyl methacrylate) in toluene (SAQ 4.6)

Retention volume (ml)	Height (cm)
130	1.0
135	12.0
140	51.4
145	89.0
150	84.0
155	51.2
160	17.8
165	4.4

4.5 Ultracentrifugation

In a gravitational field, heavy particles settle at the bottom of a column of solution by the process of *sedimentation* and this process may be exploited to determine polymer molecular weights [5]. Sedimentation is usually slow, but can be accelerated by replacing the gravitational field with a centrifugal field, making use of an *ultracentrifuge*. Such an approach involves attaching samples to a rotor that can be spun at high speed. The solute particles at a particular distance from the axis of the rotor spinning at a given angular velocity will experience a centrifugal force. The concentrations of the sample at different radii are normally monitored by measuring the refractive indices. The boundary between the solvent and the solute results in a distinct difference in the refractive indices, with the latter being measured by using Schlieren optics.

The molecular weight of a polymer in solution can be determined by examining the rate of sedimentation in such an experiment. First, the *sedimentation constant* (S) for the polymer system needs to be determined. This parameter relates the angular velocity (ω), the drift speed (s) and the radius (r), according to the following relationship:

$$S = s/r\omega^2 \tag{4.17}$$

As the speed is simply the rate of change of distance with time (t), we can write:

$$s = \mathrm{d}r/\mathrm{d}t = r\omega^2 S \tag{4.18}$$

Integration then gives the following:

$$\ln(r/r_0) = \omega^2 St \tag{4.19}$$

Thus, a plot of $\ln(r/r_0)$ versus t will be linear and S can be obtained from the slope.

Once the sedimentation constant has been determined, the molecular weight of the polymer can be calculated if the *diffusion coefficient* (D) is known. The latter measures the rate at which the polymer molecules spread down the concentration gradient, as follows:

$$M = SRT/(1 - \rho v_s)D \tag{4.20}$$

where M is the molecular weight of the sample, R is the universal (molar gas) constant, T is the temperature, ρ is the solution density and v_s is the specific volume of the solute (the volume per unit mass).

DQ 4.2

How can ultracentrifugation be used to distinguish between a styrene–isoprene copolymer and a mixture of polystyrene and polyisoprene?

Answer

The two homopolymers will show two distinct refractive index peaks, while for the copolymer only one peak will be observed.

SAQ 4.7

The rate of sedimentation of a polymer in an ultracentrifuge was monitored at 20°C, using a rotor speed of 50 000 rpm and the results obtained are listed below in Table 4.5. Calculate the molecular weight of the polymer on the basis that its partial specific volume is 0.728 $cm^3 g^{-1}$ and its diffusion coefficient is 7.62×10^{-7} $cm^2 s^{-1}$ at 20°C, with the density of the solution being 0.9981 g cm^{-3}.

Table 4.5 Sedimentation results obtained by ultracentrifugation of an unknown polymer (SAQ 4.7)

Time (s)	0	600	1200	1800
Radius (cm)	6.127	6.179	6.232	6.284

4.6 Osmometry

Osmometry is the determination of molecular weight by the measurement of osmotic pressure [6]. This technique is based upon the phenomenon of osmosis, i.e. the passage of pure solvent into a solution of sample material where the two systems are separated by a semi-permeable membrane. A typical apparatus used for this is illustrated schematically in Figure 4.7. The osmotic pressure (π) is the pressure that must be applied to the solution to stop the influx of solvent. The

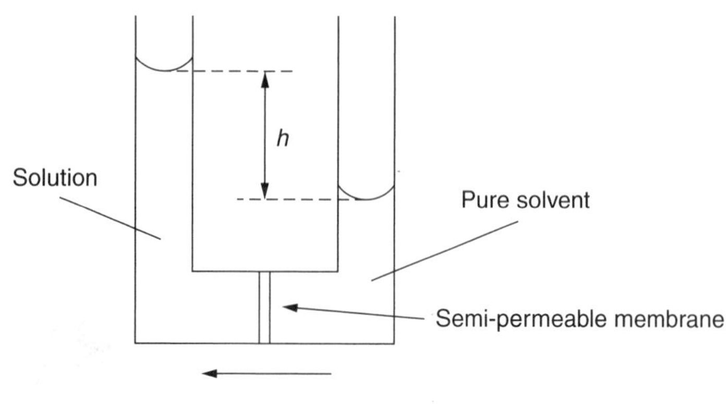

Figure 4.7 Schematic of a typical set-up used to measure osmotic pressure.

relationship between molecular weight and osmotic pressure may be derived by applying the van't Hoff equation, and is given as follows:

$$\pi/c = (RT/\overline{M}_{\mathrm{n}})(1 + Bc/\overline{M}_{\mathrm{n}} + \cdots, \text{etc.}) \tag{4.21}$$

where c is the mass concentration of the polymer, $\overline{M}_{\mathrm{n}}$ is the number-average molecular weight of the polymer, R is the universal (molar gas) constant, T is the temperature and B is the *osmotic virial coefficient*; B is a constant incorporated into the equation in order to take account of the non-ideality of the polymer solution. By plotting π/c versus c, $\overline{M}_{\mathrm{n}}$ can be obtained from the y-intercept of this plot. Alternatively, h/c versus c may be plotted, where h is the solution height measured directly from the osmometer, as the latter parameter is proportional to π .

SAQ 4.8

The osmotic pressures of solutions of poly(vinyl chloride) in cyclohexane at 298 K are presented below in Table 4.6. The density of the polymer solution is 0.980 g cm^{-3}. Determine the molecular weight of the polymer.

Table 4.6 Osmometry data obtained for various solutions of poly(vinyl chloride) in cyclohexane (SAQ 4.8)

Concentration (g l^{-1})	1.00	2.00	4.00	7.00	9.00
Height (cm)	0.28	0.71	2.01	5.10	8.00

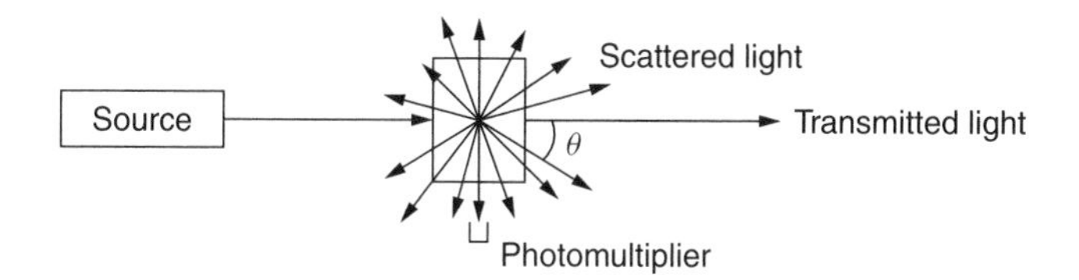

Figure 4.8 Schematic of a typical set-up used to carry out light scattering measurements.

4.7 Light Scattering

Light scattering techniques are amenable to the study of polymers because of the large size of such molecules [2, 7]. The process of *light scattering* results from light incident upon matter, as follows. An incident beam of radiation upon a sample induces vibrations in the nuclei and excitation of the electrons, with the emitted radiation being scattered in all directions. Figure 4.8 shows a typical layout of an apparatus used to carry out light scattering experiments. The photomultiplier shown can be rotated through a series of angles.

Light scattering can be used to determine the molecular weight of a polymer by employing the following relationship:

$$Hc/R(\theta) = (1/RT)(\mathrm{d}\pi/\mathrm{d}c)_T \tag{4.22}$$

where π is the osmotic pressure, θ is the angle of the incident radiation, c is the polymer concentration, H is an optical constant for a particular polymer and solvent, and $R(\theta)$ is the Rayleigh ratio, where the latter relates the intensities of the incident and the scattered light. The *Zimm*-derived equations relating the light scattering intensity to the molecular weight are given as follows:

$$(Hc/R(\theta))_{\theta=0} = (1/\overline{M}_\mathrm{w}) + 2A_2c + \cdots, \text{etc.} \tag{4.23}$$

$$(Hc/R(\theta))_{c=0} = (1/\overline{M}_\mathrm{w})[1 + 1/3(4\pi/\lambda)^2 R_\mathrm{g}^2 \sin^2(\theta/2) + \cdots, \text{etc.}] \tag{4.24}$$

These equations may be combined to construct a *Zimm plot*, as illustrated in Figure 4.9. The analysis of light scattering data can be simplified by plotting two separate graphs. First, $Hc/R(\theta)$ can be plotted as a function of c for several angles and the intercepts of the lines with the y-axis for each angle can be obtained (Figure 4.10(a)). These intercepts can then be plotted versus $\sin^2(\theta/2)$, from which the intercept of this particular plot with the y-axis gives $1/\overline{M}_\mathrm{w}$ (Figure 4.10(b)).

DQ 4.3

Why is the weight-average molecular weight, rather than the number-average value, obtained from light scattering measurements?

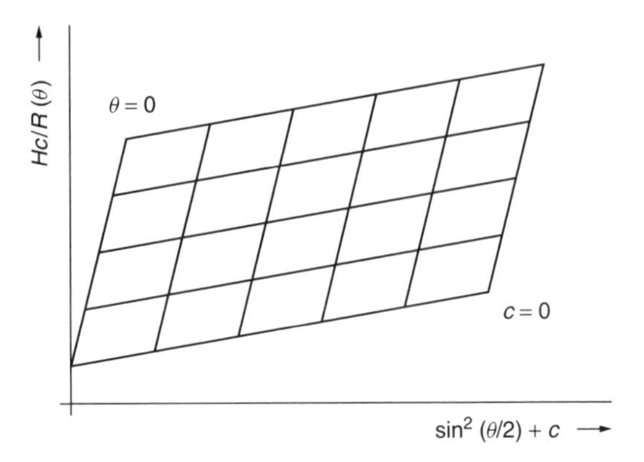

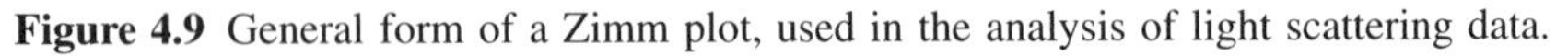

Figure 4.9 General form of a Zimm plot, used in the analysis of light scattering data.

Answer

The intensity of scattering is greater for larger particles, so the average is weighted more heavily in their favour.

SAQ 4.9

The light scattering data shown below in Table 4.7 give values of the quantity '$10^7 Hc/R(\theta)$' for solutions of an unknown polymer. Calculate the molecular weight of this material.

Table 4.7 Light scattering data obtained for various solutions of an unknown polymers (SAQ 4.9)

Concentration ($g\ l^{-1}$)	Scattering angle (°)		
	45	32	17
0.88	69.8	49.0	33.0
0.64	66.0	45.5	29.4
0.43	62.8	42.1	25.9

4.8 End-Group Analysis

The ends of polymer chains often consist of functional groups different from the monomer(s) which constitute the polymer. *End-group analysis* involves the determination of the number of moles of end-groups of a particular type in a given mass of polymer. For this technique to be applicable to a particular polymer system, the end-groups on the molecules must be amenable to quantitative analysis, usually carried out by employing titrimetry or spectroscopy [8, 9].

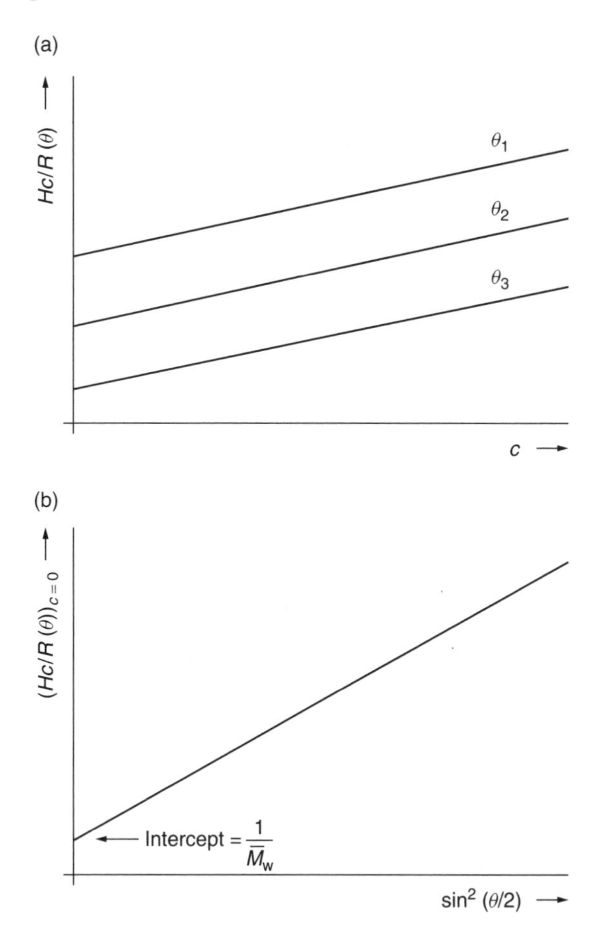

Figure 4.10 Light scattering plots, used in the determination of the weight-average molecular weight of a polymer.

Linear condensation polymers are most suited for end-group analysis because of the certainty that all chains are terminated by reactive end-groups. For polyamides, there are different end-groups possible depending on the method of preparation. For a polyamide such as nylon 6, which is a linear molecule, there is a carbonyl group at one end of the polymer chain and an amino group at the other:

$$HOOC-(CH_2)-_5\!\left[NH-CO-(CH_2)_5\right]_n NH_2$$

In this case, there is one functional group of each kind per molecule.

For linear polymers, the determination of the ends-groups gives a measurement of $\overline{M}_n$. When employing a straightforward acid–base titration, the molecular weight can be calculated by using the following equation:

$$\overline{M}_n = (\text{number of groups which can be determined per polymer molecule} \times \text{mass of polymer sample})/\text{number of moles of H}^+\text{reacted with polymer} \quad (4.25)$$

The technique tends to be restricted to low-molecular-weight linear polymers. The usual limit of molecular weight is of the order of 20 000 to 30 000, with the sensitivity of the method decreasing as the molecular weight of the polymer increases. Sources of end-groups that have not been taken into account in the end-group reaction become more significant as the molecular weight increases. In addition, the fraction of such groups becomes too small to measure with precision with increasing molecular weight.

SAQ 4.10

Nylon 6,10 was polymerized by using an excess of hexamethylenediamine. A 2.04 g sample of this nylon was dissolved in a phenol/methanol solution and titrated against a 0.0100 M HCl solution, using a thymol blue indicator. A titre of 24.50 ml HCl was recorded. Estimate the molecular weight of the nylon.

4.9 Turbidimetric Titration

In a turbidimetric titration, a polymer is dissolved in a solvent and is then slowly precipitated by the addition of a non-solvent to the solution [9]. The turbidity due to the precipitated polymer is measured either by the decrease in intensity of a beam of transmitted light or by an increase in intensity of a beam of scattered light. The solubility of a polymer is inversely proportional to the molecular weight, and thus this fractionation technique can be used to measure the relative molecular weight distribution.

When titrating with a non-solvent, the turbidity (t) must be corrected for the volume increase due to the addition of non-solvent (V_n) to the solvent (V_s), by using the following expression:

$$t = t_{obs}(V_s + V_n)/V_s \quad (4.27)$$

A distribution curve is then obtained by plotting t versus the volume fraction of the non-solvent (V_f), as follows:

$$V_f = V_n/(V_s + V_n) \quad (4.28)$$

Turbidimetric titration is an approximate method and thus can only provide comparative information about similar polymers. However, this approach does have the advantage that once the appropriate conditions have been established, it can be carried out rapidly with only small samples – particularly useful for quality control purposes.

Summary

This chapter introduced the various approaches that can be used to calculate the average molecular weights of polymers. A number of common methods used to determine such molecular weights were described. These included viscometry, size-exclusion chromatography, ultracentrifugation, osmometry, light scattering, end-group analysis and turbidimetric titration.

References

1. Lovell, P. A., 'Dilute Solution Viscometry', in *Comprehensive Polymer Science*, Vol. 1, Booth, C. and Price, C. (Eds), Pergamon Press, Oxford, UK, 1989, pp. 173–198.
2. Sandler, S. R., Karo, W., Bonesteel, J. and Pearce, E. M., *Polymer Synthesis and Characterization: A Laboratory Manual*, Academic Press, San Diego, 1998.
3. Painter, P. C. and Coleman, M. M., *Fundamentals of Polymer Science: An Introductory Text*, Technomic Publishing, Lancaster, PA, 1997.
4. Dawkins, J. V., 'Size Exclusion Chromatography', in *Comprehensive Polymer Science*, Vol. 1, Booth, C. and Price, C. (Eds), Pergamon Press, Oxford, UK, 1989, pp. 231–258.
5. Budd, P. M., 'Sedimentation and Diffusion', in *Comprehensive Polymer Science*, Vol. 1, Booth, C. and Price, C. (Eds), Pergamon Press, Oxford, UK, 1989, pp. 199–214.
6. Kamide, K., 'Colligative Properties', in *Comprehensive Polymer Science*, Vol. 1, Booth, C. and Price, C. (Eds), Pergamon Press, Oxford, UK, 1989, pp. 65–102.
7. Katime, I. A. and Quintana, J. R., 'Scattering Properties: Light and X-Rays', in *Comprehensive Polymer Science*, Vol. 1, Booth, C. and Price, C. (Eds), Pergamon Press, Oxford, UK, 1989, pp. 103–132.
8. Crompton, T. R., *Analysis of Polymers: An Introduction*, Pergamon Press, Oxford, UK, 1989.
9. Schroder, E., Muller, G. and Arndt, K. F., *Polymer Characterization*, Hanser Publishers, Munich, 1989.

Chapter 5
Structure

Learning Objectives

- To recognize the isomeric forms of polymers.
- To understand the chain dimensions of polymers.
- To understand the crystallinity of polymers.
- To appreciate orientation in polymers.
- To understand the characteristics of polymer blends.
- To appreciate the factors which can affect the thermal properties of polymers.
- To use dilatometry, infrared spectroscopy, Raman spectroscopy, NMR spectroscopy, differential scanning calorimetry, thermal mechanical analysis, dynamic mechanical analysis, optical microscopy, transmission electron microscopy, X-ray diffraction and neutron scattering to characterize the structural properties of polymers.

5.1 Introduction

Not only are the physical properties of polymers affected by molecular weight, but also by the structure of the molecular chains. Depending on the nature of the polymerization process, certain polymers are able to form different configurational isomers. Polymers are also able to form a range of conformations depending upon the backbone structure. One of the more fundamental aspects of thermoplastic polymers is their ability to form crystalline states and the presence of such regions has a significant effect on the properties of these materials. The thermal properties of thermoplastics are also important factors to be considered when deciding on the most suitable material for a particular application. The orientation by mechanical deformation and the blending of polymers are additional factors that can change the properties of polymer systems and these are also discussed in this present chapter. There are a range of techniques that are commonly used to study the structural properties of polymers. Some of the more common approaches, including dilatometry, infrared spectroscopy, Raman spectroscopy, NMR spectroscopy, thermal analysis, optical and electron microscopies, X-ray diffraction and neutron scattering, are outlined in the following sections.

5.2 Isomerism

For some polymers, polymerization can result in different configurations. For instance, when a radical attacks an asymmetric vinyl monomer, two types of addition are possible. The monomer unit in the chain may either be *head-to-head* or *head-to-tail*, as illustrated in Figure 5.1. In most polymers, the head-to-tail configuration predominates because in the head-to-head configuration steric hindrance, due to the presence of large side-groups, can occur.

In *stereoisomers*, the atoms are linked in the same order in a head-to-tail configuration, but differ in their spatial arrangement. Especially for asymmetric monomers, the orientation of each monomer adding to the growing chain is described by the term *tacticity*. There are three stereoisomers observed for vinyl polymers, namely *isotactic*, where all of the the substituent (R) groups (see Figure 5.2) are situated on the same side of the polymer chain, *syndiotactic*, where the substituent groups are located on alternate sides of the polymer chain, and *atactic*, where the substituent groups are randomly positioned along the polymer chain. Commercial PP is essentially isotactic and, due to its regular structure, is crystalline and so gives rise to good mechanical properties. In contrast, atactic PP is unable to crystallize because of its irregular structure and is a soft amorphous material that has no useful mechanical properties.

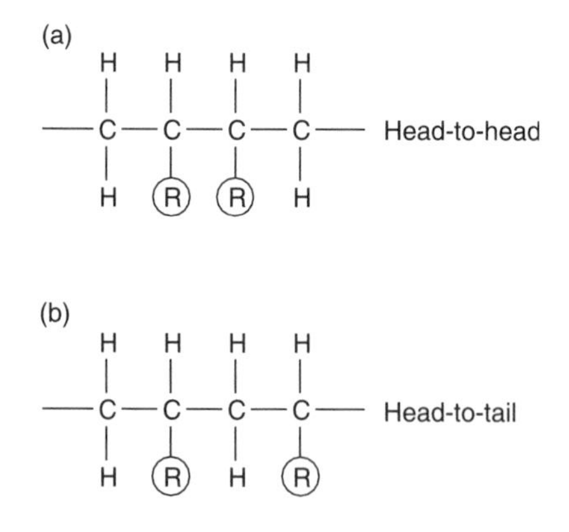

Figure 5.1 Different configurations that can exist, for example, in vinyl polymers: (a) head-to-head; (b) head-to-tail.

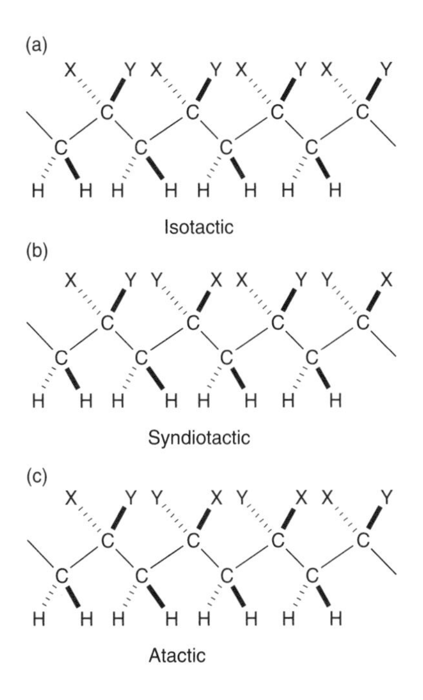

Figure 5.2 The three forms of stereoisomers observed for vinyl polymers: (a) isotactic; (b) syndiotactic; (c) atactic.

Geometrical isomerism exists in polymers which retain double bonds in the main chain. Such isomers are configurational in nature as rotation cannot take place about the double bond. Chain growth from monomers such as conjugated dienes can proceed in a number of ways, as illustrated in Figure 5.3, by using the example of isoprene. Addition takes place either via a 1,2-addition or a 3,4-addition, both of which can lead to the formation of isotactic, syndiotactic or atactic stereoisomers, or via a 1,4-addition that can result in *cis–trans* isomers. The properties of *cis*- and *trans*-isomers are often quite different. Polyisoprene provides a good example of this behaviour, where the *cis*-form of this polymer is elastomeric (natural rubber), while the *trans*-form of polyisoprene, known as *gutta percha*, is a hard, inelastic material.

DQ 5.1

Identify all of the possible isomers that can result from the addition polymerization of butadiene.

Answer

The addition polymerization of butadiene can result in the formation of poly(1,2-butadiene), poly(1,4-butadiene) and poly(3,4-butadiene).

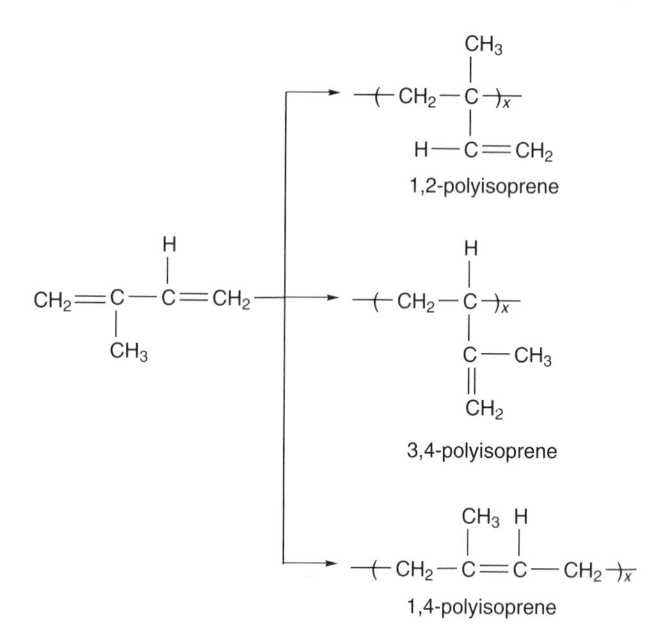

Figure 5.3 The different geometrical isomers that can occur in polyisoprene.

Poly(1,4-butadiene) can form cis- *and* trans-*isomers because of the double bond in its backbone chain. Poly(1,2-butadiene) and poly(3,4-butadiene) result in the same structures and can both form atactic, isotactic and syndiotactic isomers.*

5.3 Chain Dimensions

Polymer molecules in solution typically form random coils due to the ease of rotation around the backbone bonds. A generalized view of the polymer in such a conformation is shown in Figure 5.4. The size of a polymer molecule may be expressed in three ways. The *contour length* (R_c) is the distance from the beginning to the end of the molecule along the covalent bonds of the backbone. The *root-mean-square end-to-end distance* (R_{rms}) is the average distance between the first and last segment (as illustrated in Figure 5.5). This distance can be determined if the number of bonds is known, as follows:

$$R_{rms} = N^{1/2} l \tag{5.1}$$

where N is the number of bonds and l is the bond length. A further parameter is the *radius of gyration* (R_g). This distance is the root-mean-square distance of the elements of the chain from its centre of gravity and may be calculated by

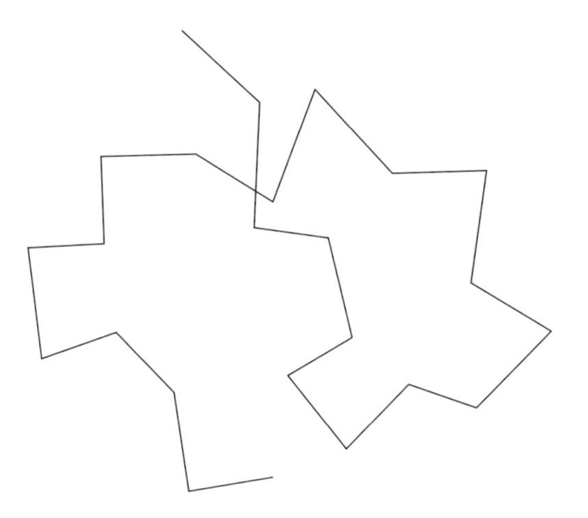

Figure 5.4 A generalized view of the random coil conformation adopted by a polymer.

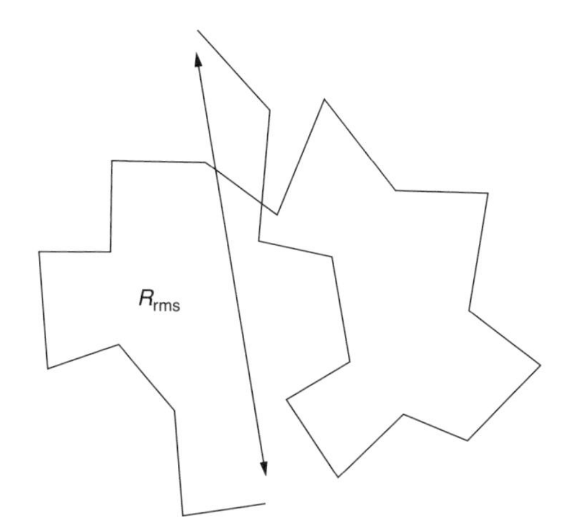

Figure 5.5 Representation of the root-mean-square end-to-end distance in a polymer molecule.

using the following equation:

$$R_g = N^{1/2}l/\sqrt{6} \tag{5.2}$$

SAQ 5.1

Calculate the contour length and the root-mean-square end-to-end distance for a polyethylene sample with a molecular weight of 280 kg mol^{-1}. Assume that the length of a C–C bond is 154 pm.

It should be noted that the bonds in a polymer chain cannot make all angles to each other as this would be a physical impossibility. There are interactions that affect the dimensions of polymer chains. These include *short-range* interactions between neighbouring atoms or groups, such as steric repulsion caused by overlapping electron clouds. There are also *long-range* interactions, such as the attractive and repulsive forces between widely separated segments that occasionally approach one another during movement. For a homoatomic chain, the distance between the chain ends can be rewritten as follows:

$$R_{\mathrm{rms}} = N^{1/2}l[(1 - \cos\theta)/(1 + \cos\theta)]^{1/2} \tag{5.3}$$

For example, in the case of PE, $\theta = 109.5^\circ$, $R_{\mathrm{rms}} = \sqrt{2}N^{1/2}l$, and $R_{\mathrm{g}} = N^{1/2}l/\sqrt{3}$. Steric repulsions also impose restrictions to bond rotation and thus:

$$R_{\mathrm{rms}} = N^{1/2}l[(1 - \cos\theta)(1 - \cos\phi)/(1 + \cos\theta)(1 + \cos\phi)]^{1/2} \tag{5.4}$$

where $\cos\phi$ is the average of the cosine of the angle of rotation of the bonds in the backbone chain.

5.4 Crystallinity

Polymer crystallinity can be described as the packing of the molecular chains so as to produce an ordered array [1]. Such *crystalline* regions of a polymer are illustrated in Figure 5.6. The remaining randomly ordered regions of a polymer

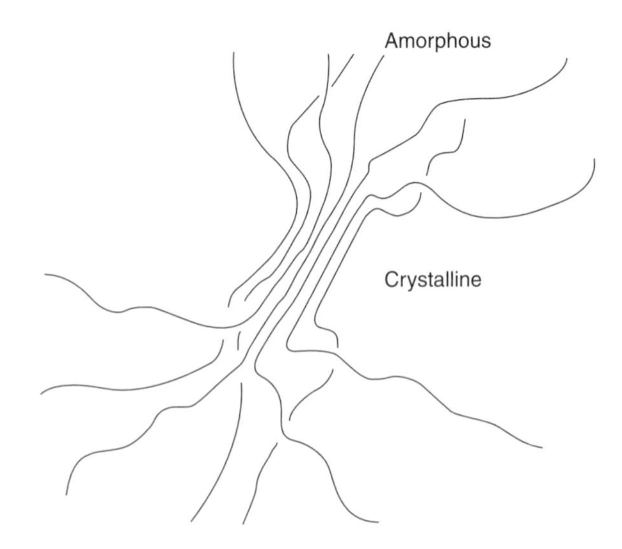

Figure 5.6 Crystalline and amorphous regions of a polymer.

are described as *amorphous* (also illustrated in Figure 5.6). Polymers are often partially crystalline (*semicrystalline*), with crystalline regions dispersed within the remaining amorphous material. The degree of crystallinity usually varies over the range 0 to 95%. The density of a crystalline polymer is greater than that of the corresponding amorphous polymer, with the degree of crystallinity being determined from density measurements, as follows:

$$\%\text{crystallinity} = 100 \times \rho_c(\rho_s - \rho_a)/\rho_s(\rho_c - \rho_a) \tag{5.5}$$

where ρ_s is the density of the sample, ρ_a is the density of the totally amorphous polymer, and ρ_c is the density of the perfectly crystalline polymer.

SAQ 5.2

The density and associated crystallinity data for two polypropylene samples are listed below in Table 5.1. Determine the density of a sample having 74.6% crystallinity.

Table 5.1 Density and crystallinity data for two polypropylene samples (SAQ 5.2)

ρ (g cm^{-3})	Crystallinity (%)
0.904	62.8
0.895	54.4

When investigating the conformations of single polymer chains, the most common low-energy conformation is the *planar zigzag* conformation. Polyethylene (PE) is the most widely studied crystalline polymer and the planar zigzag molecules in this material pack into an orthorhombic unit cell, as illustrated in Figure 5.7. The molecules are held in position by van der Waals bonding between the chain segments. The unit cell of PE has dimensions of $a = 7.41$ Å and $b = 4.94$ Å, with successive pendant chain atoms being a distance of 2.55 Å apart along the chain axis. Isotactic polymers, in which the pendant group is bulky, can adopt a *helical* conformation. A representation of a 3/1 helix formed by isotactic vinyl polymers is shown in Figure 5.8.

There are two main theories available for describing polymer crystallization. The first model developed to describe crystallization in polymers is known as the *fringed micelle model*. This model predicts that crystallization occurs by bundles coming together in regular segments from different molecules. The polymer forms a closely packed crystalline array at localized points, as illustrated in Figure 5.9. In the *lamellae model* (or *chain-folded model)*, crystallization takes place by single molecules folding themselves at intervals of about 10 nm to form lamellae which are of the order of 100 Å in thickness (Figure 5.10.). There are

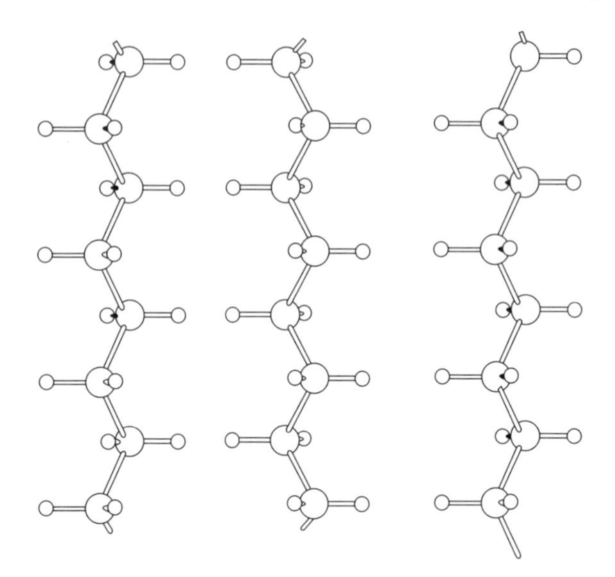

Figure 5.7 The planar zigzag conformation exhibited, by crystalline polyethylene.

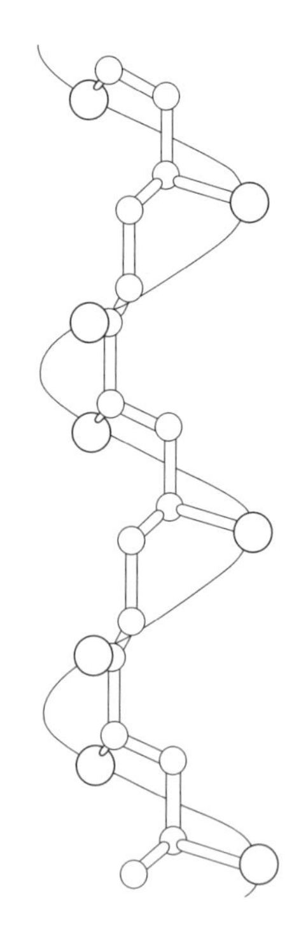

Figure 5.8 Representation of a 3/1 helix formed by isotactic vinyl polymers.

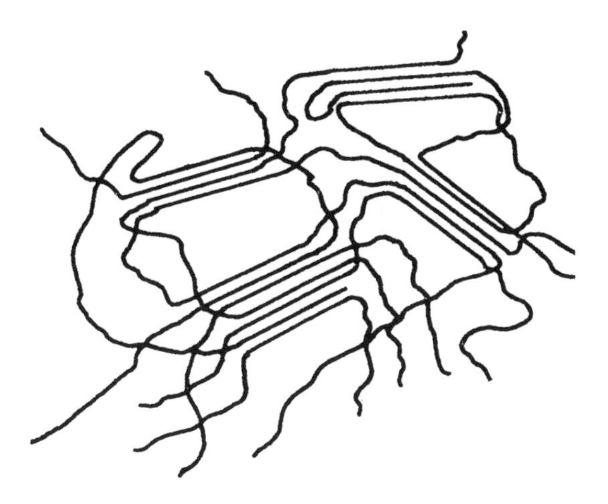

Figure 5.9 Illustration of the fringed micelle model of polymer crystallinity.

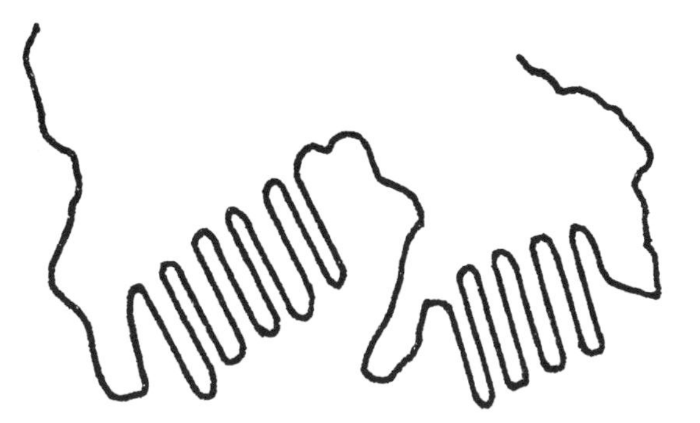

Figure 5.10 Illustration of the lamellae model of polymer crystallinity.

several types of folding suggested for polymer single crystals (see Figure 5.11.), namely, *regular*, where adjacent re-entry with sharp folds is observed, *irregular*, where adjacent re-entry is observed with loose folds, and *switchboard* which displays random re-entry. The aggregates of lamellae can also form spherulites, as illustrated in Figure 5.12.

Crystallization in polymers occurs in several stages. First, there is nucleation, which involves the formation of small particles (nuclei) of the new phase which are capable of growing. For polymers, this involves the ordering of chains in a parallel array. The second stage involves the growth of the crystalline region. The size of the crystals depends on the rate of addition of other chains to the nucleus. Spherulites grow from heterogeneous nuclei by a two-stage process. *Primary crystallization* occurs when the lamellae develop radially outward from the nucleus until stopped by interactions with the neighbouring spherulites. This

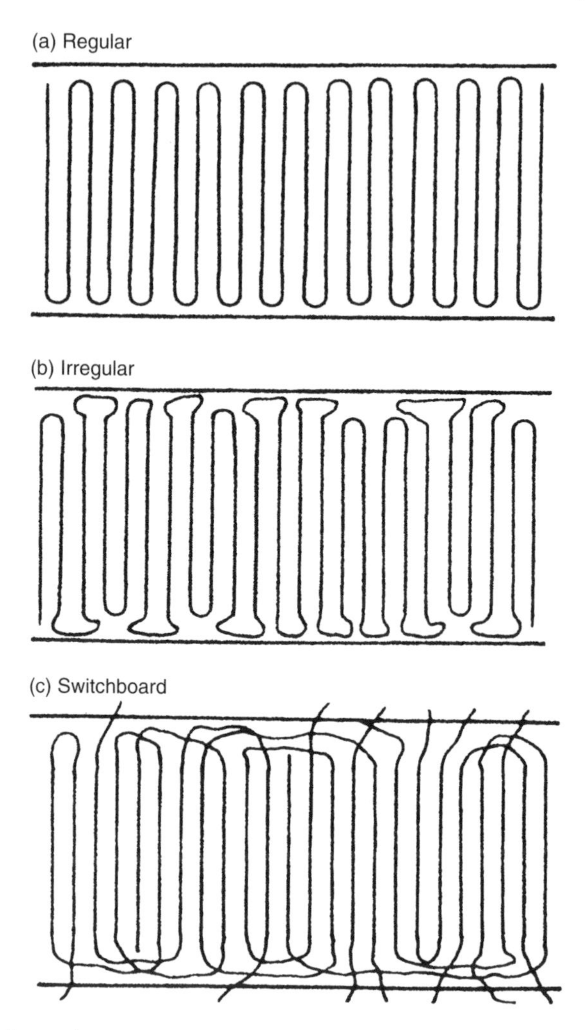

Figure 5.11 The various types of chain folding exhibited by polymer single crystals: (a) regular; (b) irregular; (c) switchboard.

can be followed by *secondary crystallization*, which proceeds with each spherulite transforming part of the remaining amorphous material.

The process of crystal growth is a thermally activated process which can be mathematically described by the following expression:

$$\nu/\nu_0 = \exp(-E_D/RT)\exp[-CT_m/T(T_m - T)] \tag{5.6}$$

where ν is the linear spherulite growth rate, ν_0 is a universal constant for crystalline polymers ($\nu_0 \sim 7.5 \times 10^8$ μm s^{-1}), E_D is the activation energy for the

Figure 5.12 Illustration of the spherulite structure formed by aggregation of polymer lamellae.

transport of polymer chains, C is an experimental constant, R is the universal (molar gas) constant, T is the temperature at which crystallization takes place and T_m is the melting temperature of the polymer.

The *Avrami equation* describes the crystallization kinetics of polymers at a fixed temperature, and is given as follows:

$$w_t/w_0 = \exp(-kt^n) \tag{5.7}$$

where t is the crystallization time in s, k is the rate constant of crystallization in s^{-n}, w_0 is the mass of the polymer melt at zero time, w_t is the mass of the melt remaining after time t, and n is the *Avrami exponent*. The latter parameter is an integer that can provide information about the geometric form of the growth. The values of n are 2, 3 or 4 for one-, two- or three-dimensional growth geometries, respectively. For example, fibril, disc or spherulite growth can be approximated by using this approach. Note that the Avrami equation can describe some, but not all, polymer systems.

Liquid crystal polymers (LCPs) demonstrate a liquid crystalline phase called the *mesophase*, which is intermediate between a solid and a liquid [2]. In the mesophase, molecules show liquid-like long-range behaviour, essentially disordered, but with some crystal-like short-range behaviour. *Lyotropic* LCPs are polymers that exhibit liquid crystalline behaviour in solution, while *thermotropic* LCPs exhibit liquid crystalline behaviour in the melt. There are different ordered structures formed by LCPs, such as *smectic*, *nematic* and *cholesteric* (Figure 5.13). In a smectic (from the Greek for 'soapy') phase, the molecule align themselves in

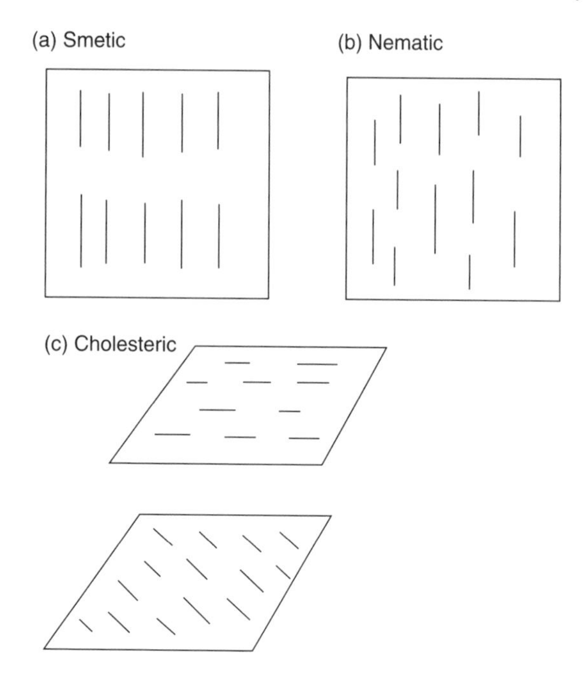

Figure 5.13 Various types of ordered structures formed by liquid crystal polymers: (a) smectic; (b) nematic; (c) cholesteric.

layers. In a nematic (from the Greek for 'thread') phase, the molecules lack the layered structure, but retain a parallel alignment. Nematic liquid crystals are used in data displays because of this behaviour and their response to electric fields. In the cholesteric phase, the molecules lie in sheets at angles that change slightly between neighbouring sheets and form helical structures. The pitch of the helices, and thus the manner in which they diffract light, changes with temperature and so these liquid crystals have colours that are temperature-dependent. They are so-named because some derivatives of cholesterol form this phase.

5.5 Orientation

The strength of polymers can be improved by molecular orientation [3]. Such orientation can be achieved by drawing, stretching or rolling a polymer. Mechanical deformation of this type results in an increased alignment of the polymer chains in the direction of the deforming force. The *degree of orientation* can be calculated by using the following equation:

$$F = \frac{1}{2}(3\cos^2\theta - 1) \tag{5.8}$$

where $\cos^2\theta$ is the mean-square cosine, averaged over all of the molecules, of the angle between a given crystal axis and a reference direction. The crystal axis is usually the chain axis, while the reference direction is the fibre axis. F values of 1, 0 and $-1/2$ describe systems with perfect, random and perpendicular alignment of the polymer chains relative to the reference direction, respectively.

Birefringence occurs as polymer chains are aligned in the process of orientation and is a useful property to monitor the latter. This is an optical phenomenon in which a polymer shows different refractive indices for plane polarized light in two perpendicular directions. Orientation birefringence is the result of a physical ordering of optically anisotropic elements, such as chemical bonds, along some preferential direction (Figure 5.14). *Anisotropy* is the observation of different physical properties in different directions. The birefringence (Δn) can be determined by using the following equation:

$$\Delta n = n_{\parallel} - n_{\perp} \tag{5.9}$$

where $n_{\parallel}$ is the parallel refractive index, and $n_{\perp}$ is the perpendicular refractive index of the polymer. Amorphous polymers display no birefringence, while semicrystalline and oriented polymers are capable of showing this behaviour.

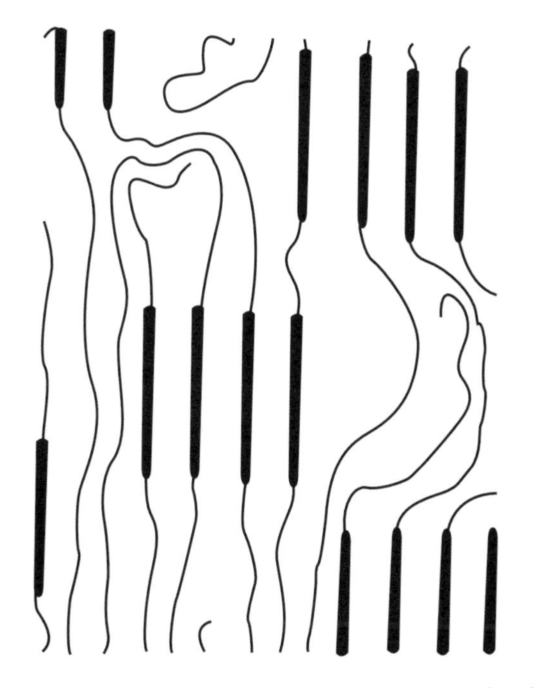

Figure 5.14 Illustration of orientation, i.e. the physical ordering of optically anisotropic elements in a polymer molecule (such as chemical bonds), which gives rise to the phenomenon of birefringence.

5.6 Blends

The majority of polymer–polymer mixtures are two-phase systems and are usually referred to as *immiscible* polymer blends [4]. However, single-phase or *miscible* blends are also possible. The physical properties and applications of blended polymers depend to a large extent on the degree of miscibility of the blend components. The mixing characteristics of two chemical components is governed by the Gibbs free energy change which occurs on their mixing (ΔG_m), according to the following equation:

$$\Delta G_m = \Delta H_m - T\Delta S_m \tag{5.10}$$

where T is the temperature, and ΔH_m and ΔS_m are the enthalpy and entropy of mixing, respectively. If ΔG_m is negative, mixing may occur spontaneously and a solution will be produced as a result. In the case of low-molecular-weight compounds, ΔS_m is highly positive, and hence $-T\Delta S_m$ will be negative and mixing will be particularly favoured. Usually, ΔH_m, which depends upon energetic interactions between the molecules, is positive and therefore not favourable for mixing. The positive ΔH_m is often outweighed by a negative entropy term so that mixing will then occur. The more similar the chemical nature of the two components, then the lower is the value of ΔH_m and the more likely mixing becomes. In polymer–polymer mixtures, ΔS_m is much less positive as randomization is quite restricted because of the large size of the molecules involved. Although the value of the $-T\Delta S_m$ term is still favourable, it is much less so and in most cases is unable to overcome an unfavourable ΔH_m term. Thus, most polymers are immiscible and when mixed together they form a two-phase rather than a single-phase system, with one polymer dispersed as particles or domains in a matrix of the other polymer. In summary, for two polymers to be miscible, the Gibbs free energy of mixing must be negative. Miscibility of polymers that are attracted to each other is thermodynamically favoured. Miscibility is also more likely in blends containing low-molecular-weight polymers.

DQ 5.2

Why do polystyrene (PS) and poly(phenylene oxide) (PPO) form a miscible blend?

Answer

There is hydrogen bonding between the PS aromatic rings and the methyl groups in PPO. As these polymers are attracted to each other, miscibility is thus thermodynamically favoured.

DQ 5.3

Why do poly(ether ether ketone) (PEEK) and polytetrafluoroethylene (PTFE) form an immiscible blend?

Answer

There is no potential for the formation of intermolecular interactions between the side-groups of these molecules and miscibility of these materials is therefore not thermodynamically favoured.

The majority of polymer blends are immiscible and have the advantage that properties may be more readily tailored. There are a number of phase morphologies exhibited by such blends. The different types of dispersions of polymers in the matrix of an immiscible polymer are illustrated in Figure 5.15. Blends may consist of one phase dispersed as simple spheres in a matrix of the other polymer. The dispersed phase may also take the form of platelets or fibrils. A morphology consisting of an interpenetrating network of phases is also feasible. An *interpenetrating polymer network* (IPN) consists of an assembly of at least two polymers in network form, one of which is prepared or cross-linked in the presence of the other. The properties of immiscible polymer blends can be improved by the addition of a *compatibilizer*. Compatibilization is the process of modification of the interfacial properties of an immiscible polymer blend to improve the adhesion and blend properties. This process can involve the incorporation of block or graft copolymers which are identical to those in the respective phases.

5.7 Thermal Behaviour

The glass transition temperature (T_g) is the temperature at which an amorphous polymer ceases to be brittle and glassy and becomes less rigid and rubbery [5]. As a polymer is heated up to the T_g, the molecular rotation about single bonds become significantly easier. The change of volume for a polymer with increasing temperature is shown in Figure 5.16, while values of the T_g for some common polymers are listed in Table 5.2.

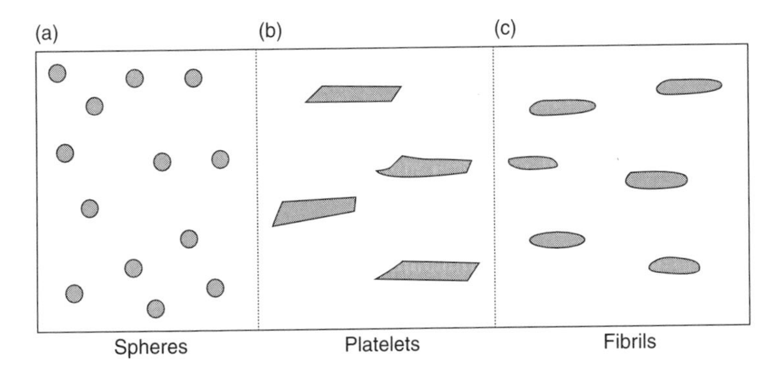

Figure 5.15 The different morphologies exhibited by immiscible blends of polymers: (a) spheres; (b) platelets; (c) fibrils.

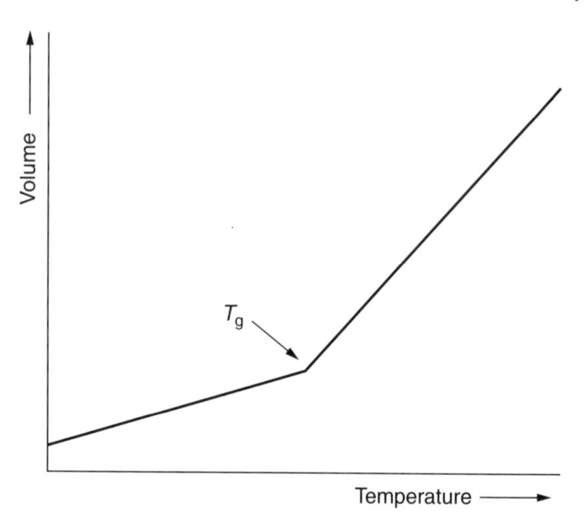

Figure 5.16 The volume change for a polymer with increasing temperature.

Table 5.2 T_g values for some common polymers

Polymer	T_g (°C)
PDMS	−123
LDPE	−110
PTFE	−97
HDPE	−90
Polychloroprene	−50
PVDC	−17
PP	−15
PMMA	12
PVA	29
Nylon 6,6	57
PET	69
PPS	85
PVC	87
PPO	90
PS	100
PAN	104
PEEK	143
PC	150
PI	280
Polyaramid	375

The melting point of a crystalline polymer (T_m) can be characterized. Crystalline polymers melt over a range of temperature rather than exhibiting a sharp melting point. This is due to the mixture of amorphous and crystalline phases that is present. For crystalline polymers, there are some useful empirical rules that may be used to describe the thermal properties. For example, there is a relationship between T_g and T_m for crystalline polymers, depending on the polymer structure. For unsymmetrical polymers (e.g. PVC), a ratio of $T_g/T_m \sim 0.66$ may be used, while for symmetrical polymers (e.g. PVDC), a ratio of $T_g/T_m \sim 0.5$ is applicable (where T_g and T_m are both measured in degrees kelvin).

For polymers in the amorphous state, for a temperature region on either side of the T_g there are three well-defined molecular relaxation processes (movement of molecular segments). The α relaxation is observed above the T_g and results from large-scale conformational rearrangements of the polymer chain backbone. The β relaxation is observed just below the T_g and results from hindered rotations of the side-groups independent of the polymer chain backbone. Finally, the γ relaxation is observed well below the T_g and is associated with the disordered regions of the polymer.

There are a number of factors which influence the glass transition temperature. First, the microstructure of the polymer is fundamental to the value of T_g. The presence of pendant groups, such as side-chains or branches, attached to the polymer backbone increases the energy required to rotate the molecule about the primary bonds in the main chain and so causes steric hindrance. For example, PP has a T_g of 5°C, while PE has a T_g of −20°C. This difference in temperature may be attributed to the presence of the methyl group in PP. The side group in the latter inhibits the freedom of rotation of the polymer chain, while PE has a more flexible backbone free of any inhibiting side-groups.

Chain flexibility also affects the T_g and is determined by the ease with which rotation occurs around the primary bonds in the backbone chain. The presence of rigid structures, such as phenyl groups in the backbone, will increase the T_g. Halogenated polymers also have a much 'stiffer' backbone and so display higher T_g values. For example, PTFE has a T_g of 115°C, which is considerably higher than that of PE ($T_g = -20$°C).

DQ 5.4

Determine which of poly(vinyl alcohol) (PVAl) and poly(ethylene oxide) (PEO) would possess the highest glass transition temperature.

Answer

The OH side group of PVAl makes rotation difficult and so this polymer possesses the higher T_g *of the pair (= 85°C). PEO has a flexible main chain and therefore chain rotation is easier. This polymer has a* T_g *of −41°C.*

The presence of intermolecular bonds, such as hydrogen bonds, in polymers leads to higher T_g values. Cross-linking reduces the free volume available and so increases the T_g.

The tacticity of polymers also has an important influence on the T_g value. The latter varies, according to stereochemistry, in the following order: syndiotactic > atactic > isotactic. This trend may be illustrated by using poly(methyl methacrylate) (PMMA) as an example, where isotactic, atactic and syndiotactic PMMA have T_g values of 38, 105 and 120°C, respectively.

Molecular weight is also a factor which must be taken into account when determining the T_g. Higher-molecular-weight polymers have less ease of movement than low-molecular-weight materials. PS is a good illustration of this dependence, e.g. PS with a molecular weight of 3000 shows a T_g of 43°C, while a PS sample of molecular weight 300 000 has a T_g of 99°C.

The presence of plasticizers significantly affects the T_g of polymers. *Plasticizers* essentially act as spacers between polymer molecules and disrupt the secondary bonding. The T_g of a plasticized polymer may be determined by using the following relationship:

$$1/T_g = w_s/(T_g)_s + w_l/(T_g)_l \tag{5.11}$$

where T_g is the glass transition temperature for the plasticized polymer, $(T_g)_s$ is the glass transition temperature for the pure polymer, $(T_g)_l$ is the glass transition temperature for the pure plasticizer, w_s is the weight fraction of the polymer (the solid) and w_l is the weight fraction of the plasticizer (the liquid), where all temperatures are measured in degrees Kelvin.

SAQ 5.3

When 100 g of PVC is mixed with 25 g of a plasticizer, the T_g of the polymer is lowered from 87 to 0°C. What T_g is to be expected from a mixture of 100 g of PVC with 100 g of the plasticizer?

Copolymers show a glass transition temperature resulting from the thermal properties of the component polymers. Equation (5.12) below can be used to predict the T_g of a random copolymer by using the mass fractions of the respective monomers from which the copolymer is made:

$$1/(T_g)_{AB} = w_A/(T_g)_A + w_B/(T_g)_B \tag{5.12}$$

where $(T_g)_A$ is the glass transition temperature of polymer component A, $(T_g)_B$ is the glass transition temperature of polymer component B, $(T_g)_{AB}$ is the glass transition temperature of the AB copolymer, and w_A and w_B are the weight fractions of components A and B, respectively (where all temperatures are measured in degrees Kelvin).

SAQ 5.4

A styrene–butadiene copolymer is found to contain 47% styrene and 53% butadiene (by weight). The T_gs of styrene and butadiene are 112 and −78°C, respectively. Determine the T_g of this copolymer.

Immiscible polymer blends show separate T_gs associated with each phase. Certain polymer blends of compatible homopolymers show a single T_g value, thus indicating miscibility. The glass transition is related to the segmental motion of 50–100 backbone chain atoms, that is, a domain diameter of 2–3 nm. The T_g is dependent on the composition and there are two commonly used relationships to describe the glass transitions of blends. The first of these, i.e. the *Gordon–Taylor equation*, is given by the following:

$$\Sigma w_i \Delta C_{pi}(T_{gi} - T_g) = 0 \tag{5.13}$$

where w_i is the weight fraction of component i, ΔC_{pi} is the heat capacity (at constant pressure) of the glass transition of component i, T_{gi} is the glass transition temperature of component i and T_g is the glass transition temperature of the blend. For a two-component system, the equation is as follows:

$$w_1(T_{g1} - T_g) + kw_2(T_{g2} - T_g) = 0 \tag{5.14}$$

where the variable k is the ratio of the heat capacities of the glass transition temperatures of the polymer components, given as follows:

$$k = \Delta C_{p2}/\Delta C_{p1} \tag{5.15}$$

In the above, k can be used as an empirical parameter, and thus a measure of the miscibility. For other polymers, the *Fox equation* may be more applicable, as given by the following:

$$\Sigma w_i(1 - T_g/T_{gi}) = 0 \tag{5.16}$$

For a two-component system, the Fox equation is given as follows:

$$1/T_g = w_1/T_{g1} + w_2/T_{g2} \tag{5.17}$$

SAQ 5.5

A miscible 50/50 (wt%) blend of a polybenzimidazole (PBI) and a polycarbonate (PC) was produced. The T_gs of the PBI and PC are 425 and 150°C, respectively. Predict the T_g of this blend, assuming that the Fox equation is obeyed.

5.8 Dilatometry

The rate of polymer crystallization can be followed by recording density changes in a dilatometer [6]. In this method, the polymer is heated initially to a temperature greater than the melting temperature. When the sample is completely molten, the dilatometer is placed in a thermostat which is set at the temperature selected for crystallization to take place and is then allowed to equilibrate. The general shape of the plot of dilatometer height as a function of time obtained in such crystallization experiments is illustrated in Figure 5.17. If secondary crystallization occurs, tailing towards the end of the curve may be observed and h_∞ may be more difficult to measure.

The mass fraction, w_t/w_0, of the polymer determined by the Avrami equation (equation (5.7)) can be related to the volume changes and heights measured in the dilatometer, as follows:

$$w_t/w_0 = (V_t - V_\infty)/(V_0 - V_\infty) = (h_t - h_\infty)/(h_0 - V_\infty) = \exp(-kt^n) \quad (5.18)$$

where h_t is the height at time t, V_t is the volume at time t, h_0 is the height at time zero, V_0 is the volume at time zero, h_∞ is the height at the end of the process, and V_∞ is the volume at the end of the process. By applying logarithms to equation (5.18), a linear relationship can be obtained, as follows:

$$\log_{10}[-\ln(h_t - h_\infty)/(h_0 - h_\infty)] = n\log_{10} k + n\log_{10} t \quad (5.19)$$

Thus, a plot of $\log_{10}[-\ln(h_t - h_\infty)/(h_0 - h_\infty)]$ versus $\log_{10} t$ (known as an *Avrami plot*) gives a slope of n.

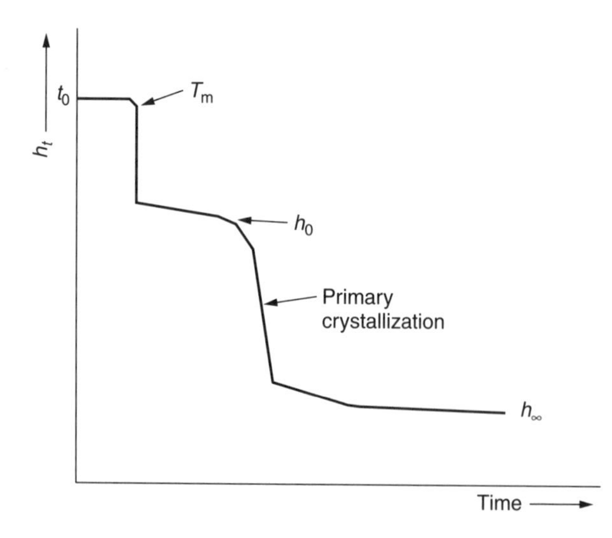

Figure 5.17 Polymer crystallization, as monitored by dilatometry.

SAQ 5.6

The crystallization of an elastomer at −5°C was studied by using a dilatometer, with the results obtained being given below in Table 5.3. The initial height was 10.00 cm and the height at the end of the process was 9.83 cm. Use an Avrami plot to determine the (Avrami) exponent for the crystallization of the elastomer at this temperature. Comment on the value that you obtain.

Table 5.3 Dilatometer data obtained for the crystallization of an elastomer at −5°C (SAQ 5.6)

Time (h)	Height (cm)
5	9.99
7	9.98
11	9.94
12	9.93
13	9.91
22	9.85
23	9.84

5.9 Infrared Spectroscopy

Infrared spectroscopy has been successfully applied to the study of the tacticity of a number of vinyl polymers [7–9]. There is a relationship between the stereoregularity and the presence of regular chains and this leads to the appearance of infrared bands due to such chains. For example, the infrared spectra of the isotactic, syndiotactic and atactic forms of PP display characteristic differences, as shown in Figure 5.18. The absorbance values at 970 and 1460 cm^{-1} do not depend upon the tacticity, whereas the absorbances at 840, 1000 and 1170 cm^{-1} are characteristic of isotactic PP, and the absorbance at 870 cm^{-1} is characteristic of syndiotactic PP. Such differences are due to the different helical structures present in the isomers and can be used to estimate the fractions of isotactic and syndiotactic sequences in particular samples.

Branching in polymers can also be examined by using infrared spectroscopy [7, 8]. The most common infrared studies of the branch contents of polymers have been carried out for PE. A comparison of the spectra of low-density polyethylene (LDPE), which is branched, and high-density polyethylene (HDPE), which is unbranched, demonstrates differences due to the presence of such branches. A band observed at 1378 cm^{-1} can be monitored to determine the methyl group concentration.

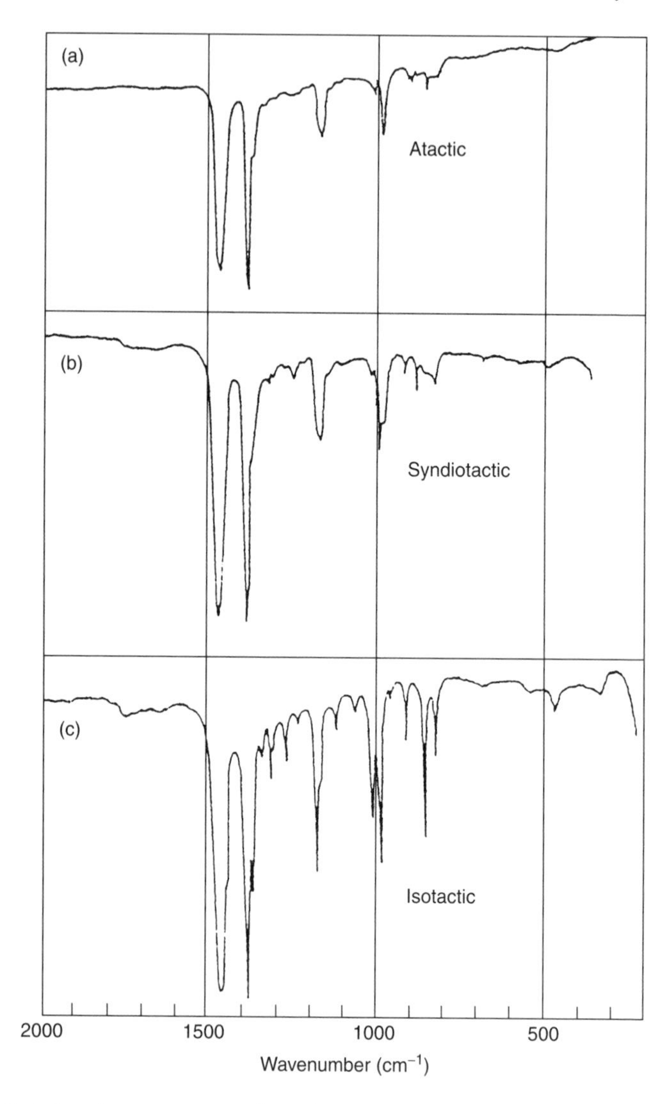

Figure 5.18 Infrared spectra showing the various stereoisomers of polypropylene: (a) atactic; (b) syndiotactic; (c) isotactic. From W. Klöpffer, *Introduction to Polymer Spectroscopy*, Springer–Verlag, Berlin (1984).

Infrared spectroscopy provides a suitable method for studying the presence of crystalline regions in polymers as the infrared modes are sensitive to changes in bond angles [7, 8]. Again, the most widely studied polymer in this case is PE. The crystalline regions of this polymer consist of molecules in a *trans*-conformation, and a 622 cm^{-1} band due to CH_2 groups is representative of a *trans–trans* structure. The amorphous regions of PE show *trans–gauche* and *gauche–gauche* structures, and methylene wagging bands at 1303, 1353 and 1369 cm^{-1} can be used to determine the concentrations of such structures.

Polyurethanes (PUs) are particularly suited to infrared investigation because they contain functional groups such as N–H, C=O and C–H. The major infrared modes common to PUs are given in Table 5.4. Such polymers are extensively hydrogen-bonded, with the proton donor being the N–H group of the urethane linkage. The hydrogen-bond acceptor may be in either the hard segment (the carbonyl of the urethane group) or the soft segment (an ester carbonyl or ester oxygen). The relative amounts of the two types of hydrogen bonds are determined by the degree of micro-phase separation. An increase in this separation favours the inter-urethane hydrogen bonds. The degree of hydrogen bonding of the N–H groups can be studied by examining the N–H stretching region of the spectrum. The presence of a shoulder in the vicinity of 3450 cm^{-1} is indicative of free N–H groups.

DQ 5.5

Hydrogen bonding plays a fundamental role in the structural and physical properties of nylons and is the most significant type of intermolecular interaction which influences the infrared spectrum of this class of polymers. The temperature-dependence of the infrared spectra of various nylons was investigated, with the infrared spectra of nylon 6 at 25 and 235°C being shown below in Figure 5.19. What do the differences in these spectra say about the structure of nylon 6 at higher temperatures?

Answer

The main changes observed in the spectra with increasing temperature are associated with the N–H and C=O stretching modes in the 3450–3300 and 1700–1600 cm^{-1} ranges, respectively. These changes are due to a breakdown of the hydrogen bonds that occur between adjacent molecules.

Table 5.4 The major infrared modes of polyurethanes

Wavenumber (cm^{-1})	Assignment
3445	N–H stretching (non-hydrogen bonded)
3320–3305	N–H stretching (hydrogen bonded)
2940	Asymmetric C–H stretching
2860	Symmetric C–H stretching
1730	C=O stretching (non-hydrogen bonded)
1710–1705	C=O stretching (hydrogen bonded)
1645–1635	C=O stretching (hydrogen-bonded urea carbonyl)

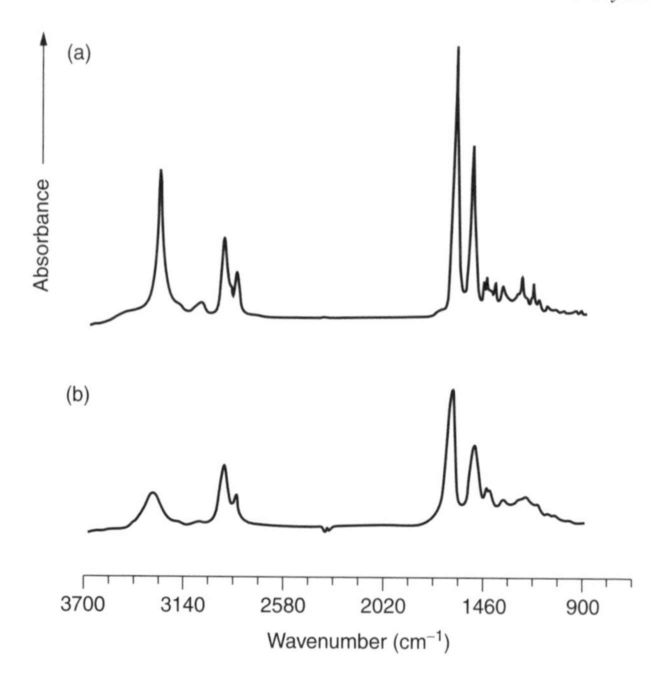

Figure 5.19 Infrared spectra of nylon 6 obtained at temperatures of (a) 25 and (b) 235°C (DQ 5.5).

Infrared spectroscopy is particularly applicable to the study of orientation in polymers. Infrared bands can often be assigned to the crystalline and amorphous regions of a polymer [7, 8]. Orientation can be observed in infrared spectroscopy because infrared absorbance is due to the interaction between the electric field vector and the molecular dipole transition moments due to the molecular vibrations. The absorbance is at a maximum when the electric field vector and the dipole transition moment are parallel to each other, and zero when the orientation is perpendicular. The orientation of molecular components can be characterized by using the dichroic ratio, which is defined as $A_{\parallel}/A_{\perp}$, where $A_{\parallel}$ is the absorbance parallel to, and $A_{\perp}$ is the absorbance perpendicular to the chain axis. The orientation of poly(ethylene terephthalate) (PET) has been well characterized by using infrared spectroscopy. Figure 5.20 illustrates the infrared spectra of oriented PET for cases where the electric field vector is parallel and perpendicular to the drawing direction. These spectra clearly show that the PET spectrum is sensitive to the orientation of the sample, as the differing alignment of the molecules results in changes in the intensities of a number of the infrared modes.

Infrared spectroscopy is now quite a commonly used technique for the examination of polymer blends [10]. If two polymers are immiscible, the resulting infrared spectrum should be the sum of the spectra of the two components.

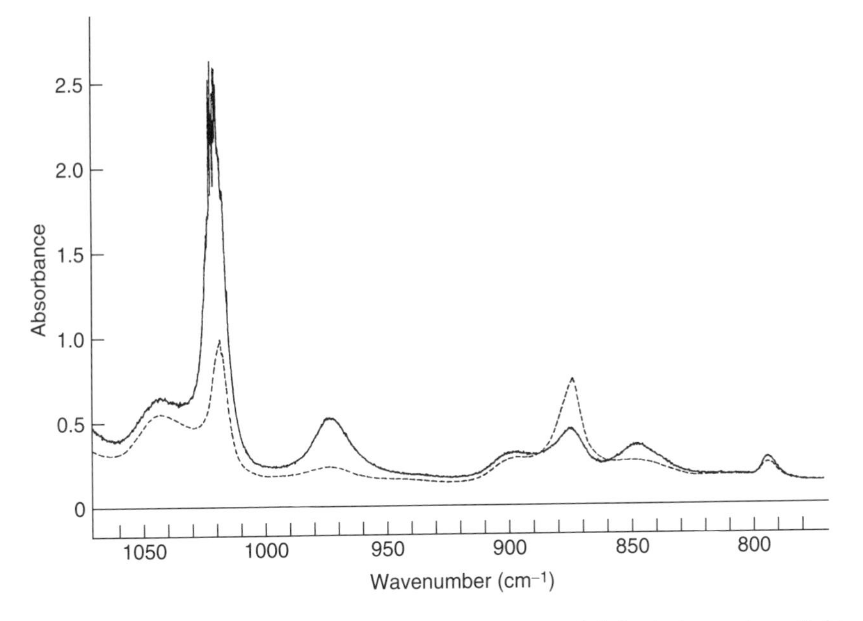

Figure 5.20 Infrared spectra of oriented poly(ethylene terephthalate): (——) parallel to the drawing direction; (- - - -) perpendicular to the drawing direction. From Bower, D. I. and Maddams, W. F., *The Vibrational Spectroscopy of Polymers*, © Cambridge University Press, 1989. Reproduced by permission of Cambridge University Press.

Phase-separation implies that the component polymers in the blend will have an environment similar to the pure polymers. If the polymers are miscible, there is then the possibility of chemical interactions between the individual polymer chains. Such interactions can lead to differences between the spectra of the polymer in the blend and those of the pure components. Generally, frequency shifts and band broadening are taken as evidence of chemical interactions between the components in a blend and are indicative of miscibility.

DQ 5.6

FTIR spectroscopy has been used to investigate the miscibility of certain poly(vinyl phenol) (PVPh) $\text{+}CH_2\text{–}CH(C_6H_6)OH\text{+}_n$ blends [11]. PVPh was blended in one case with poly(ethylene oxide) (PEO) $\text{+}CH_2\text{–}CH_2\text{–}O\text{+}_n$ and in the other case with poly(vinyl isobutyl ether) (PVIE)$\text{+}CH_2\text{–}CH(OC_4H_9)\text{+}_n$. Figures 5.21 and 5.22 below show the O–H stretching regions of the infrared spectra of these blends for a number of compositions. Are these blends miscible? If so, what is the nature of the interactions between the component polymers?

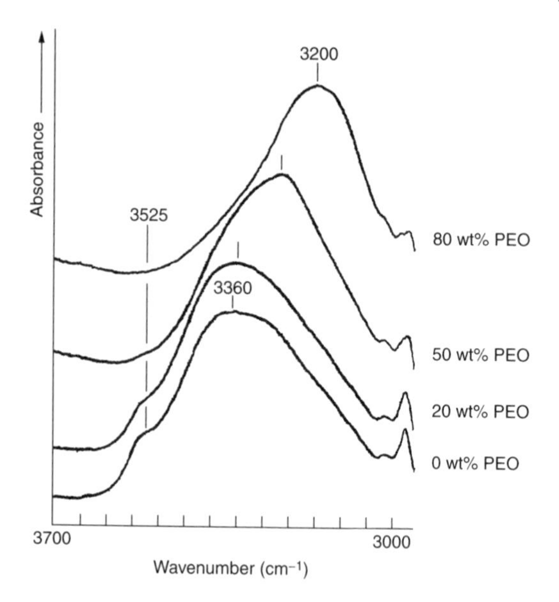

Figure 5.21 Fourier-transform infrared spectra of various blends of poly(vinyl phenol) (PVPh) and poly(ethylene oxide) (PEO) (DQ 5.6). Reprinted from *Polymer*, **26**, Moskala, E. J., Varnell, D. F. and Coleman, M. M., *Concerning the Miscibility of Poly(vinylphenol) Blends-FITR Study*, 228–234, Copyright (1985), with permission from Elsevier Science.

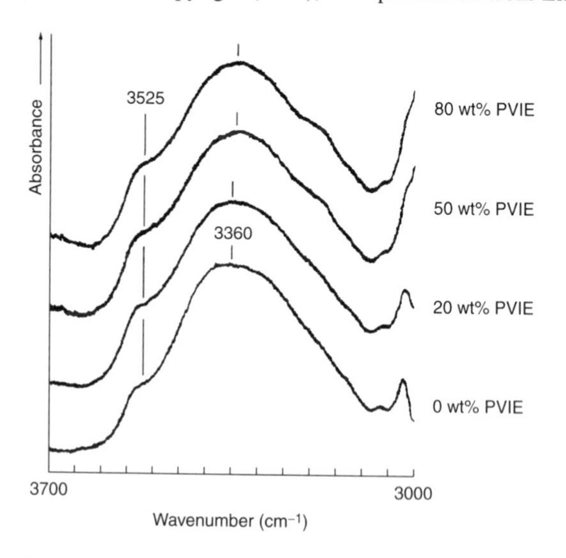

Figure 5.22 Fourier-transform infrared spectra of various blends of poly(vinyl phenol) (PVPh) and poly(vinyl isobutyl ether) (PVIE) (DQ 5.6). Reprinted from *Polymer*, **26**, Moskala, E. J., Varnell, D. F. and Coleman, M. M., *Concerning the Miscibility of Poly (vinylphenol) Blends-FITR Study*, 228–234, Copyright (1985), with permission from Elsevier Science.

Answer

Figure 5.21 shows the O–H stretching region of the PVPh component of its blend with PEO. The band at 3360 cm^{-1} shifts to a lower frequency near 3200 cm^{-1} with an increasing concentration of PEO. This change reflects an increase in hydrogen bonding between the PVPh hydroxyl groups and the PEO ether oxygens. Such an interaction indicates that the PVPh/PEO blend is miscible. Figure 5.22 illustrates the O–H stretching region of the PVPh/PVIE blend. No changes to the hydroxyl group of PVPh are observed in this case, thus indicating that there is no interaction between these two polymers, and so this blend in immiscible.

SAQ 5.7

An infrared spectroscopy study of a 50/50 (wt%) blend of a polybenzimidazole (PBI) and a polycarborate (PC) was carried out. The percentage of the total carbonyl absorbance due to the component mode at 1751 cm^{-1} as a function of temperature for the blend is shown below in Figure 5.23. Show that this blend is miscible (assume that the Fox equation holds), and describe the nature of the intermolecular bonding which is present in the system.

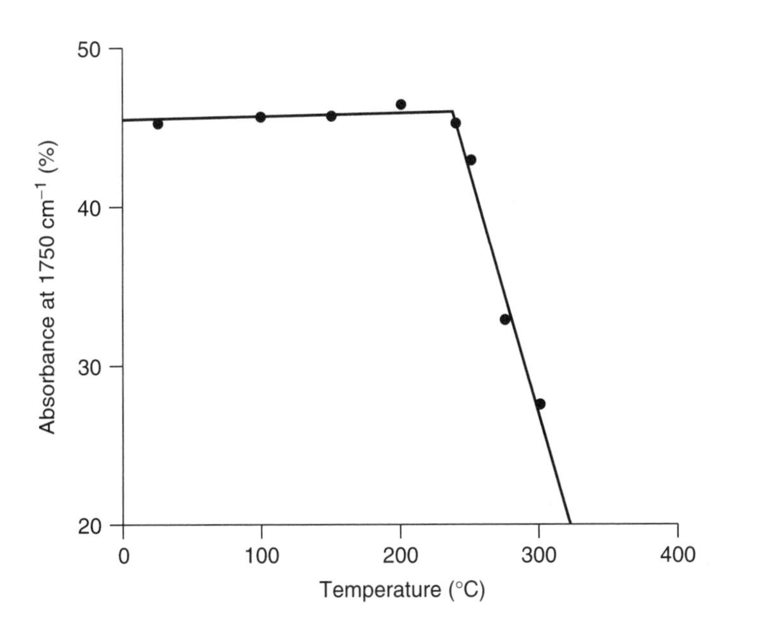

Figure 5.23 Variation of the infrared carbonyl mode of a 50/50 (wt%) blend of a polybenzimidazole (PBI) and a polycarbonate (PC) as a function of temperature (SAQ 5.7).

5.10 Raman Spectroscopy

In Raman spectroscopy, the C=C stretching mode is easily observed and also sensitive to its environment, thus making this a useful technique for the study of elastomers [12, 13]. For example, the different isomers of polybutadiene can be readily identified and quantified. Figure 5.24 illustrates the FT Raman spectra of three different polybutadiene samples. As mentioned above, the C=C stretching band is sensitive to its environment, with the *cis*-, *trans*- and vinyl C=C bands appearing at 1651, 1666 and 1639 cm^{-1}, respectively.

Raman spectroscopy has proved to be a useful spectroscopic technique for investigating the nature of the thermal changes induced in poly(ether ether ketone) (PEEK) [14]. The relative intensity of the C–O–C stretching mode has been followed as a function of temperature in order to examine the crystallization process of this polymer. This material displays two different stages during crystallization. The first of these occurs in the region of 120–130°C and is associated with the partial bond rotation of the ether linkages prior to crystallization. This is manifested by a decrease in relative intensity of the symmetric C–O–C stretching

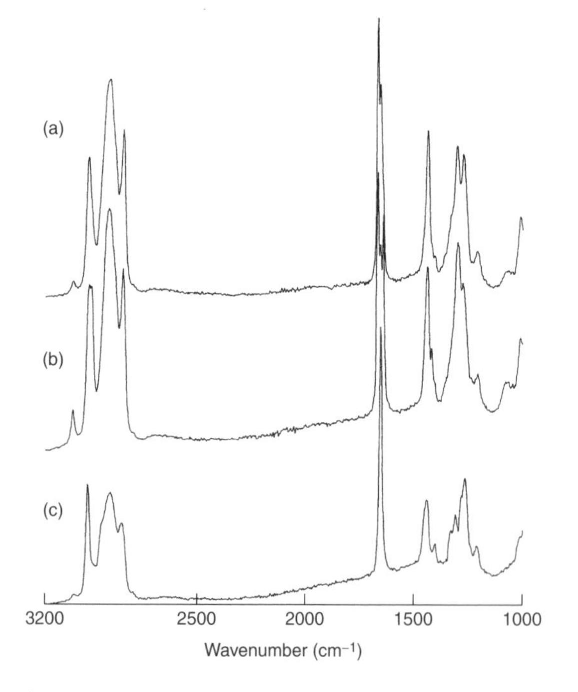

Figure 5.24 Fourier-transform Raman spectra of three different polybutadiene samples: (a) 'high-*trans*' polybutadiene ('Intene 50'); (b) 'high-vinyl' polybutadiene; (c) *cis*-polybutadiene ('Europrene'). From P. Hendra *et al*, *Fourier Transform Raman Spectroscopy: Instrumentation and Chemical Applications*, Ellis Horwood (1999).

mode. At temperatures greater than the T_g (143°C), a second process involving the motion of the whole chain occurs and is characterized by the rotation of the benzophenone linkages near the T_m (334°C). In the Raman spectrum, an increase in the intensity of the C–O–C stretching mode is observed at this temperature.

5.11 Nuclear Magnetic Resonance Spectroscopy

^1H and ^{13}C NMR spectroscopies are widely used for characterizing polymer structures. Stereochemistry, isomerism and branching, plus head-to-head and head-to-tail additions, can all be detected by using NMR spectroscopy. Branched polymers can be characterized because absorptions due to the branch points on the chains may be identified in a NMR spectrum [15]. The degree of branching can be calculated by referencing such absorptions to those nuclei in the repeat units. Figure 5.25 shows the proton-decoupled ^{13}C NMR spectra of high-density

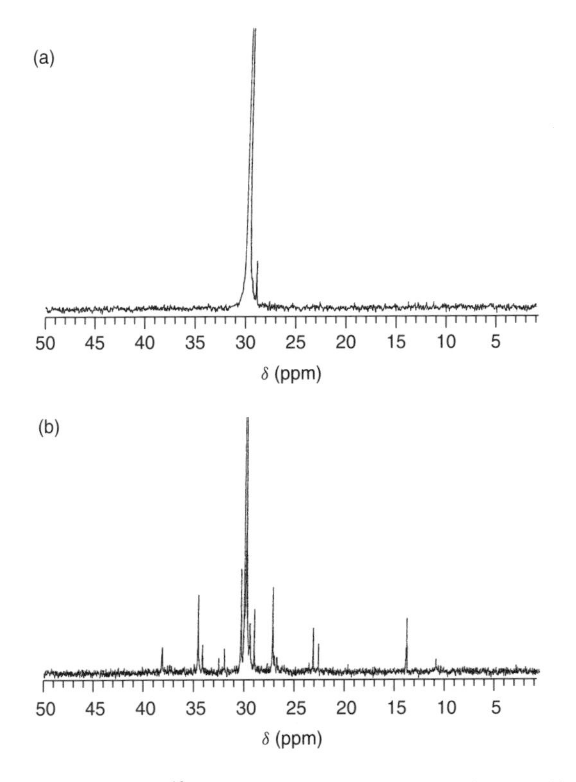

Figure 5.25 Proton-decoupled ^{13}C NMR spectra obtained for (a) high-density and (b) low-density polyethylene samples. From Sandler, S. R., Karo, W., Bonesteel, J. and Pearce, E. M., *Polymer Synthesis and Characterization: A Laboratory Manual*, © Academic Press, 1998. Reproduced by permission of Academic Press.

polyethylene (HDPE) and low-density polyethylene (LDPE) [16]. HDPE shows no branching, with only a major peak near 30 ppm. LDPE shows a branched structure, with peaks due to different branch types (butyl, amyl, hexyl, etc.), as can be observed in the figure.

Figure 5.26 illustrates the ^{1}H NMR spectra of atactic, isotactic and syndiotactic polypropylene (PP). These spectra show multiplet absorptions due to ^{1}H–^{1}H spin–spin coupling [15]. For example, the methyl absorption is split by spin–spin coupling with the C–H ^{1}H atoms and is observed as a doublet in the range $\delta = 0.8$–0.9 ppm. For comparison, Figure 5.27 illustrates the ^{13}C NMR spectrum of the same stereoisomers of PP. The ^{13}C absorptions appear as singlets because the spectra were recorded by the use of ^{1}H decoupling. The ^{13}C NMR spectra of this polymer demonstrate that the CH_2 and CH_3 absorptions are sensitive to tacticity.

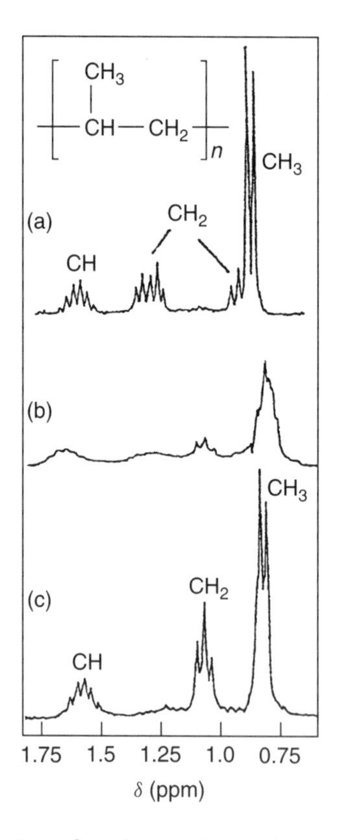

Figure 5.26 ^{1}H NMR spectra of various polypropylene stereoisomers: (a) isotactic; (b) atactic; (c) syndiotactic. From Tonelli, A. E., *NMR Spectroscopy and Polymer Microstructure*, Copyright © VCH, 1989. Reprinted by permission of John Wiley & Sons, Inc.

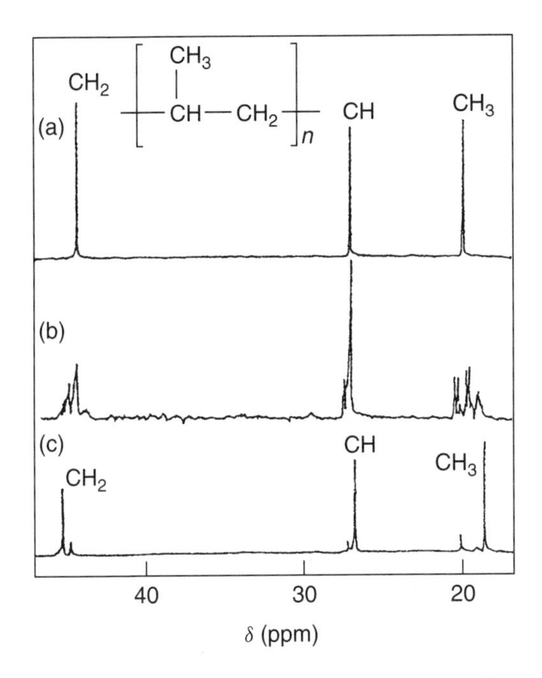

Figure 5.27 ^{13}C NMR spectra of various polypropylene stereoisomers: (a) isotactic; (b) atactic; (c) syndiotactic. From Tonelli, A. E., *NMR Spectroscopy and Polymer Microstructure*, Copyright © VCH, 1989. Reprinted by permission of John Wiley & Sons, Inc.

SAQ 5.8

Table 5.5 below summarizes the ^1H NMR absorptions for isotactic and syndiotactic poly(methyl methacrylate) (PMMA) [16]. Use this information to identify the stereoisomers of PMMA illustrated in the ^1H NMR spectra shown below in Figure 5.28 [17]. One spectrum is of predominantly syndiotactic PMMA, while the other represents isotactic PMMA.

Table 5.5 ^1H NMR absorption data obtained for isotactic and syndiotactic samples of poly(methyl methacrylate) (SAQ 5.8)

Group	δ (ppm)	
	Isotactic	Syndiotactic
Ester methyl, O–CH_3	Singlet 3.5	Singlet 3.5
Methylene, CH_2	Quartet 1.4, 1.6, 2.1, 2.3	Singlet 1.9
Backbone methyl, CH_3	1.3	1.1

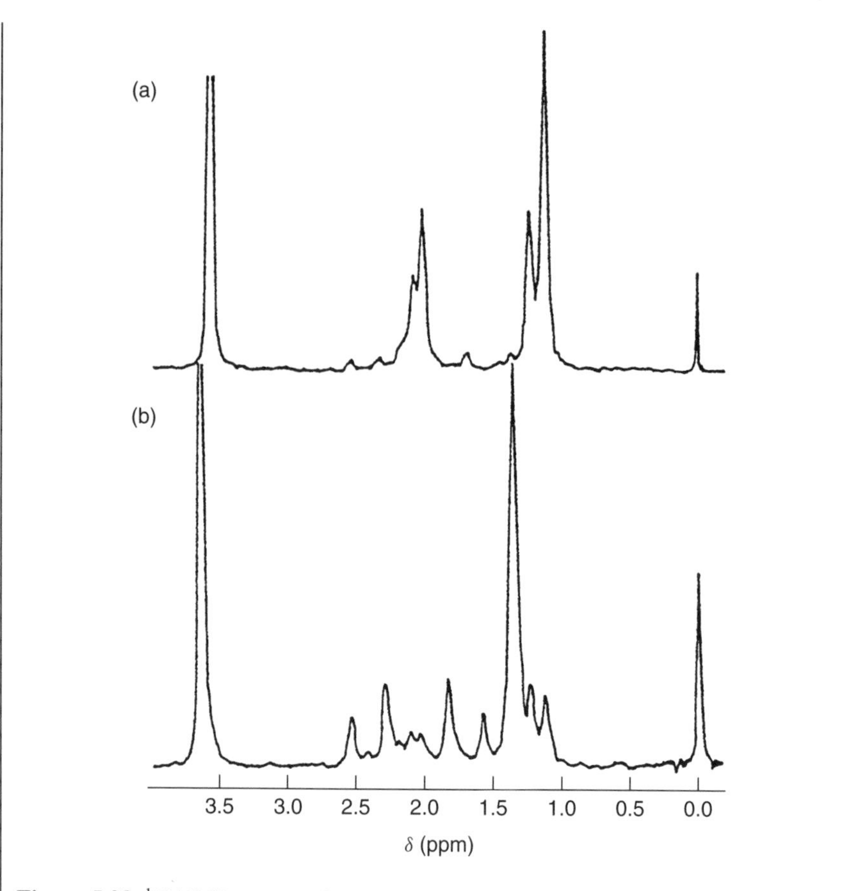

Figure 5.28 ^{1}H NMR spectra of two different stereoisomers of poly(methyl methacrylate) (SAQ 5.8). From Bovey, F. A., *High Resolution NMR of Macromolecules*, © Academic Press, 1972. Reproduced by permission of Academic Press.

5.12 Thermal Analysis

5.12.1 Differential Scanning Calorimetry

The glass transition temperature of a polymer can be detected by using differential scanning calorimetry (DSC), as an endothermic shift from the baseline is observed at the T_g in the traces of crystallizable polymers. Such a change results from an increase in heat capacity due to the increased molecular motions in the material. It should be noted that the T_g value will depend greatly on the heating and cooling rates used in a DSC run.

As the T_m of a polymer corresponds to a change from the solid to liquid state, this transition gives rise to an endothermic peak in the DSC curve. Such a peak enables the melting point and the heat of fusion to be determined by using this technique. The width of the melting peak provides an indication of the range of crystal sizes and also their perfection. Above the T_m, the polymer will degrade at the *degradation temperature* (or *decomposition temperature*) (T_d). This transition can be accompanied by either an exothermic or an endothermic peak. For crystallizable polymers, an additional transition is observed between the T_g and T_m values. This is known as the *crystallization temperature* (T_c), i.e. the temperature at which ordering and production of the crystalline regions occur. The polymer chains have sufficient mobility at this particular temperature to crystallize and an exothermic peak is then observed.

DQ 5.7

Figure 5.29 below shows the DSC curve obtained for a sample of poly(ethylene terephthalate) (PET) produced by rapid cooling of the polymer melt. Explain each of the transitions observed in this trace.

Answer

In Figure 5.29, the inflection in the region of 60–80°C corresponds to the T_g *of PET. The mobility of the PET molecules above the* T_g *leads*

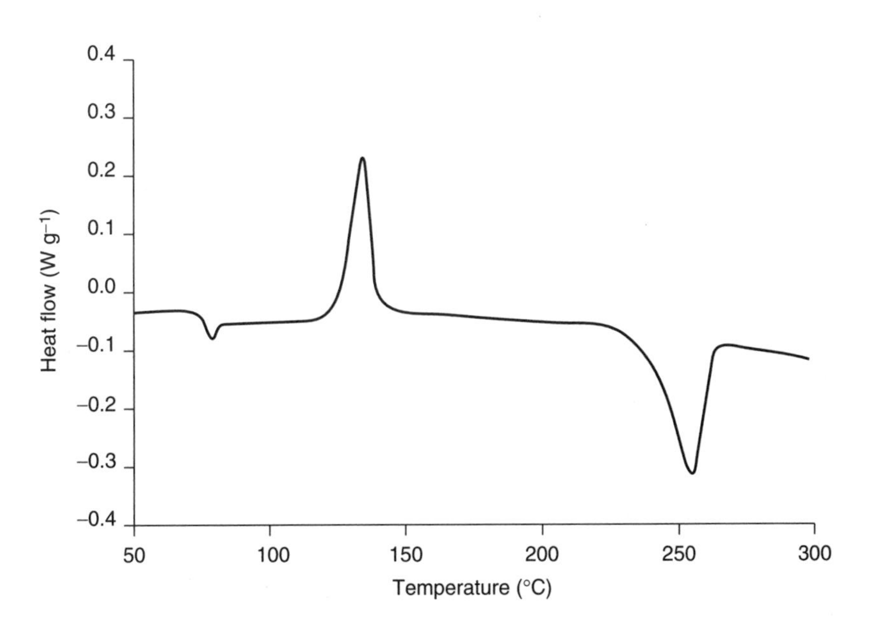

Figure 5.29 DSC curve obtained for a sample of poly(ethylene terephthalate) produced by rapid cooling from the melt (DQ 5.7).

to crystallization at $T_c = 150-180°C$*, as evidenced by the large peak in this temperature range. The peak in the range 230–260°C corresponds to the melting temperature (*T_m*).*

SAQ 5.9

DSC has been used to investigate the miscibility of poly(ether ether ketone) (PEEK) blends with both a poly(ether imide) (PEI) and a liquid crystalline polymer (LCP). The T_g values of the PEEK in both of these blends were determined for a range of concentrations, with the results being presented below in Table 5.6. Using these data, decide whether these PEEK blends are miscible.

Table 5.6 Thermal analysis data obtained for blends of poly(ether ether ketone) (PEEK) with a poly(ether imide) (PEI) and a liquid crystalline polymer (LCP) (SAQ 5.9)

PEEK (wt%)	T_g (°C)	
	PEEK/PEI	PEEK/LCP
0	219	100
25	197	148
50	180	147
75	160	147
100	145	149

Thermal analytical methods may also be used to investigate the influence of thermal history on the melting behaviour of polymers. An example is given in Figure 5.30, which shows the DTA curves of two samples of PEEK with different thermal histories, i.e. one amorphous sample quenched from 400°C, and a semicrystalline sample slow-cooled from 400°C. Note that during annealing, the samples are heated to the required temperature for a specified time under vacuum in order to bring about the morphology change.

The crystallinity of polymers can be measured using DSC by employing the following equation:

$$\% \text{ crystallinity} = (\Delta H_a - \Delta H)/(\Delta H_a - \Delta H_c) \times 100 \qquad (5.20)$$

where ΔH is the enthalpy change of the unknown specimen, ΔH_a is the enthalpy change of the pure amorphous standard and ΔH_c is the enthalpy change of the pure crystalline standard. The relationship can be simplified by assuming that $\Delta H_a = 0$, and therefore we can write:

$$\% \text{ crystallinity} = 100\Delta H/\Delta H_c \qquad (5.21)$$

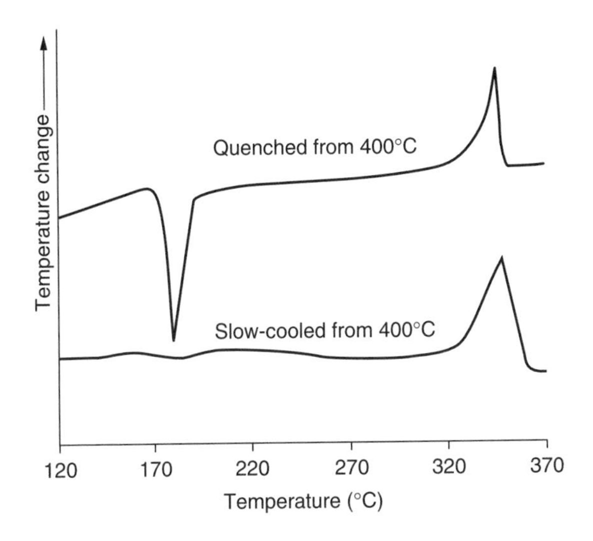

Figure 5.30 DTA curves of two samples of poly(ether ether ketone) with different thermal histories, i.e. quenched and slow-cooled from 400°C.

Alternatively, the area under the DSC melting peak of a semicrystalline polymer can be determined by comparison with a standard sample of known crystallinity. The percentage crystallinity is thus obtained from the following expression:

$$\% \text{ crystallinity} = (\text{area of sample melting peak} \times \% \text{ standard}) / (\text{area of standard melting peak}) \qquad (5.22)$$

> **SAQ 5.10**
>
> A single crystal of polyethylene (PE) weighing 9.6 mg is found to have an enthalpy of melting of 2.65 J, while a commercial sample of the same polymer weighing 6.8 mg is found to have an enthalpy of melting of 1.10 J. Calculate the percentage crystallinity of the commercial sample.

5.12.2 Thermal Mechanical Analysis

Thermal mechanical analysis (TMA) is a technique which involves the deformation of a sample under non-oscillating stress measured as a function of time or temperature [18]. Figure 5.31 illustrates schematically the components of a typical TMA apparatus. In this technique, a probe is lowered on to the surface of a sample and the force exerted on the latter by the probe is then varied. The whole assembly is enclosed in a temperature-controlled atmosphere.

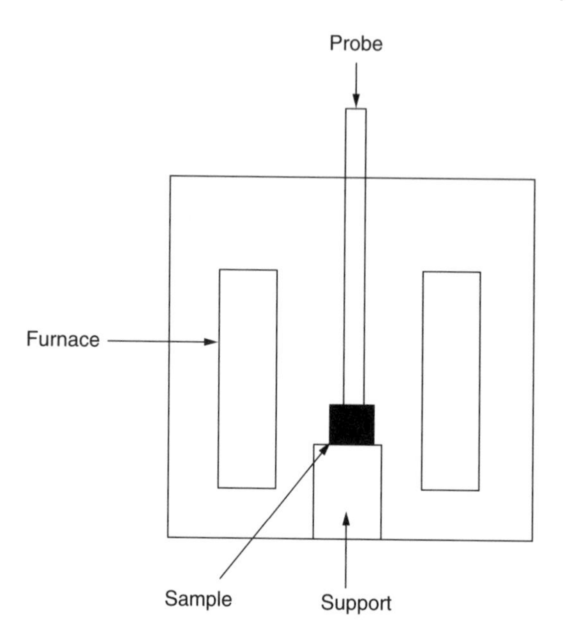

Figure 5.31 Schematic of a typical apparatus used for carrying out thermal mechanical analysis.

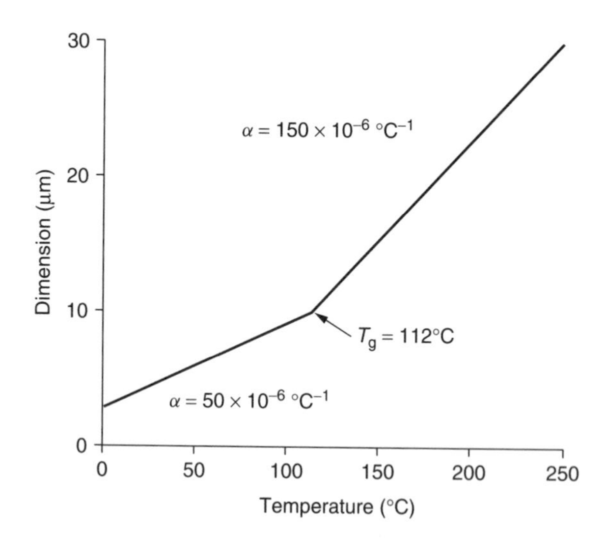

Figure 5.32 Thermal mechanical analysis results obtained for the coefficient of thermal expansion as a function of temperature (measurements made for an epoxy-based circuit board).

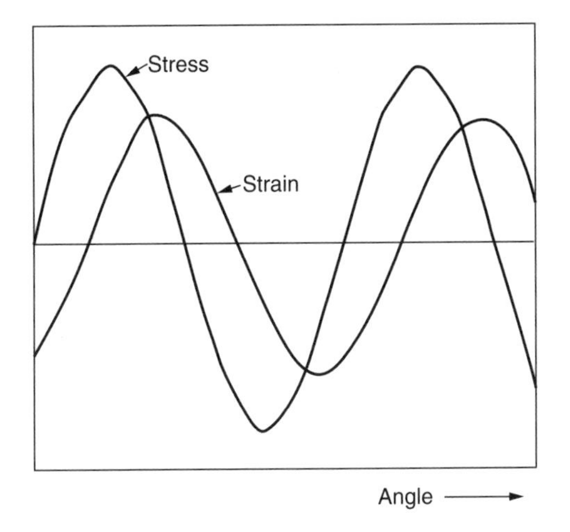

Figure 5.33 Stress–strain behaviour observed in dynamic mechanical analysis tests.

TMA can be used to determine the coefficient of thermal expansion. This latter parameter, α, between two defined temperatures is given by the following:

$$dl/dt = l_0 \alpha dT/dt \tag{5.23}$$

where l_0 is the original length of the sample, dl is the change in the sample length and dT is the temperature difference. Thus, by plotting the displacement as a function of temperature, the α value can be obtained from the slope of the plot. The value of α changes abruptly at the T_g. Figure 5.32 illustrates how TMA may be used to determine the glass transition temperature, with this figure showing the results obtained from an epoxy-based circuit board fabrication.

5.12.3 Dynamic Mechanical Analysis

Dynamic mechanical analysis (DMA) is a method which determines the elastic modulus of a material and its mechanical damping or energy dissipation characteristics as a function of frequency and temperature [16, 18]. *Viscoelasticity* will be discussed in more detail in Chapter 8, but DMA is introduced here as it provides a suitable method of characterizing the thermal transitions of polymeric systems.

DMA tests are carried out by vibrating the sample and varying the applied frequency. The changes can be related to the relaxation processes in a polymer. Generally, the stress is varied sinusoidally with time. As a result of time-dependent relaxation processes, the strain lags behind the stress, as illustrated in Figure 5.33. DMA instruments are similar in design to those used for thermal mechanical analysis. The technique can be operated in various modes, including *flexure* (measures the dynamic modulus), tension (measures the T_g), torsion (measures the T_m), shear (examines the relaxation behaviour) or compression (investigates the cross-link density).

For a frequency of oscillation $\omega/2\pi$ Hz, the stress, σ, at any given time t, is given by the following:

$$\sigma = \sigma_0 \sin \omega t \tag{5.24}$$

where σ_0 is the maximum stress. The corresponding strain, ε, is given by:

$$\varepsilon = \varepsilon_0 \sin (\omega t - \delta) \tag{5.25}$$

where δ is the phase angle, where the latter represents the amount that the strain lags behind the stress.

The stress can be resolved into two parts, namely one in-phase, and one 90° out-of-phase, with the strain. If these conditions apply to the torsion experiment, it is also possible to define two shear moduli, denoted as E' and E''. The first shear modulus (E') represents the part of the stress in-phase with the strain, divided by the strain, as follows:

$$E' = \sigma_1/\varepsilon_0 \tag{5.26}$$

The parameter E' is proportional to the recoverable energy and is called the *storage modulus*. The second shear modulus (E'') is the peak stress which is 90° out-of-phase with the strain, divided by the peak strain, and is given by the following:

$$E'' = \sigma_2/\varepsilon_0 = \sigma_1 \tan \delta/\varepsilon_0 \tag{5.27}$$

In this case, E'' is proportional to the energy dissipated as heat per cycle, and is called the *loss modulus*. The ratio of the moduli is defined as tan δ:

$$\tan \delta = E''/E' \tag{5.28}$$

The two moduli can be combined to form the *complex modulus* (E^*), as follows:

$$E^* = E' + iE'' \tag{5.29}$$

where $i = \sqrt{(-1)}$ (the complex number). These moduli may be used to provide information about the properties of polymers. For glassy polymers, E' is high. Such polymers have highly restricted structures and so exhibit poor elasticity. In these cases, no strain energy is lost as heat. For rubbery polymers, E' is low. For such polymers, there is a greater contribution from the viscous element and much strain is lost as heat.

An important dynamic technique used to study polymers is *dynamic mechanical thermal analysis* (DMTA). During such an experiment, a specimen is subjected to sinusoidal mechanical loading (stress), which induces a corresponding extension (strain) in the material. The experiment is carried out over a certain temperature range, which is typically varied between -100 and $+200°C$. It is normal to define the dynamic mechanical behaviour of a polymer in terms of E' or tan δ.

SAQ 5.11

DMTA was carried out on a rubber-filled epoxy resin. The storage modulus as a function of temperature for the sample is shown below in Figure 5.34. Calculate the value of tan δ at 135°C, given that the loss modulus at this temperature is 450 MPa.

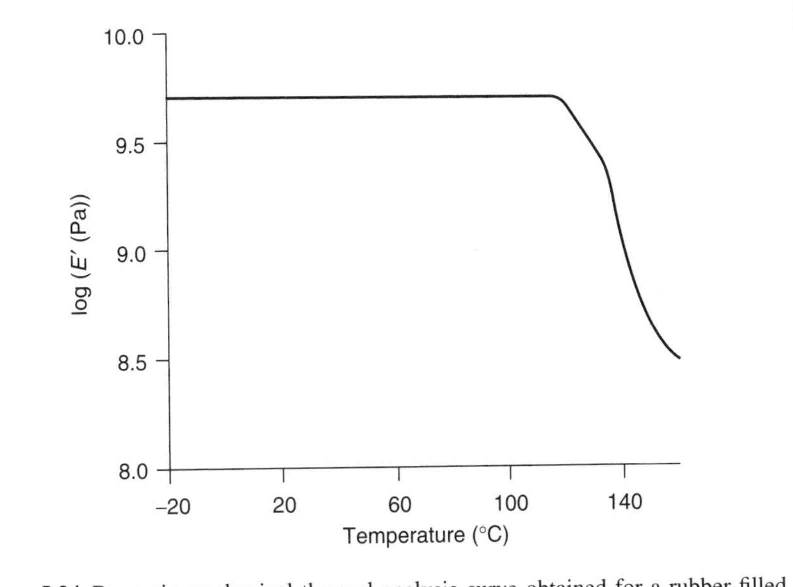

Figure 5.34 Dynamic mechanical thermal analysis curve obtained for a rubber-filled epoxy resin (SAQ 5.11).

DMA can be used to identify the relaxation processes in polymers. Figure 5.35 illustrates the transition behaviour of a semicrystalline polymer, with 'a' and 'c' referring to the amorphous and crystalline phases, respectively. Most notably, when the polymer passes through the T_g, the storage modulus often decreases by two or three orders of magnitude, and then passes through a maximum.

DQ 5.8

Figure 5.36 below shows the results of a DMA study on polychloroprene. The sample of polymer was subjected to tensile loading at a frequency of 1 Hz over a temperature range from -100 to $+20°C$. What is the T_g of the polychloroprene sample?

Answer

The maximum in the tan δ values corresponds to the T_g. *Hence, this sample shows a* T_g *of* $-32\,^{\circ}C$. *The* cis-*form of polychloroprene has a* T_g *of* $-20\,^{\circ}C$, *while the* trans-*form shows a* T_g *of* $-45\,^{\circ}C$, *thus indicating that the sample studied contains both isomeric forms of polychloroprene [19].*

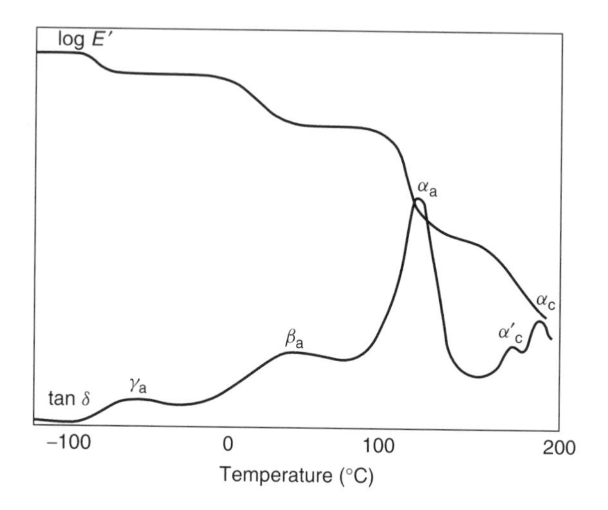

Figure 5.35 Dynamic mechanical analysis curve illustrating the transition behaviour of a semicrystalline polymer: a, amorphous phase; c, crystalline phase.

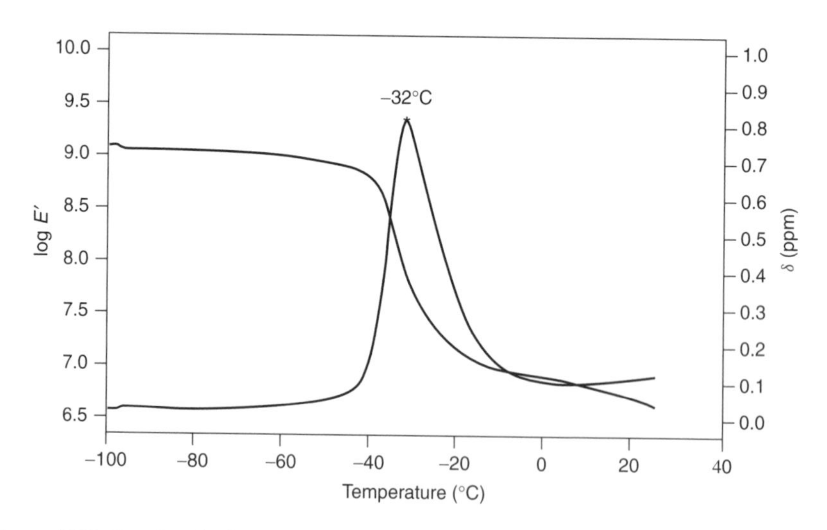

Figure 5.36 Results obtained from a dynamic mechanical analysis study of polychloroprene (DQ 5.8).

DMA also lends itself to the investigation of blend miscibility. Where a blend is miscible, a single peak representing the T_g will be observed at a value intermediate between the values of the pure component polymers. If a blend is immiscible, two distinct T_g transitions will be observed which have the same values as the component polymers.

5.13 Optical Microscopy

Microscopy can be employed to study the morphology of polymers. There are a number of experimental techniques based on the light microscope [16, 20, 21]. Figure 5.37 illustrates the various types of optical microscopy techniques which are commonly used to examine polymers. *Optical microscopy* (OM) involves the interaction of light with a sample. Magnification of the sample from $\times$ 2 to $\times$ 2000 is attainable and a resolution of the order of 0.5 μm is possible, depending on the limits of the instrument and the nature of the sample being examined.

The normal mode of operation in OM is *bright-field*. Here, the transmitted light observed in based on variations in density and colour within the sample. This mode is usually not suitable for pure polymer samples, but can be used to detect the presence of additives, such as pigments and fillers. However, a *dark-field* mode may be used to improve the contrast. In this mode, any undiffracted or unscattered light is excluded from the image-forming process and the background light intensity is near zero.

Interference microscopy can be employed to determine the dimensions or roughness of samples. In this technique, the illumination is split into two beams and, in reflection, one beam is reflected from the sample while the other is reflected from a reference mirror. The resulting interference pattern can be used to measure the thickness of the sample or the sample roughness in reflection. The plane of vibration of a beam emerging from the sample depends on the phase-difference, δ, between the interfering beam and is measured by the angle between the optical axes of the plate and the analyser, α, as follows:

$$2\alpha = \delta(2\pi/\lambda)(n_2 - n_1)l \tag{5.30}$$

where λ is the wavelength of the light, l is the sample thickness, and n_1 and n_2 are the refractive indices of the sample and the surrounding medium, respectively. Thus, if the refractive indices are known, l can then be calculated.

Polarized light microscopy may be used to study oriented samples such as fibres and liquid crystalline polymers. This technique involves the study of samples by using polarized light. A standard microscope can be fitted with polarizing filters. In the extinction position, the orientation direction is aligned parallel to one of the polarizers direction, at either 0 or 90°, with the maximum intensity being observed at the 45° position. Polarizing microscopes can also be used to observe the *small-angle light scattering* (SALS) patterns which are produced when polarized light passes through a sample having a spherulitic structure.

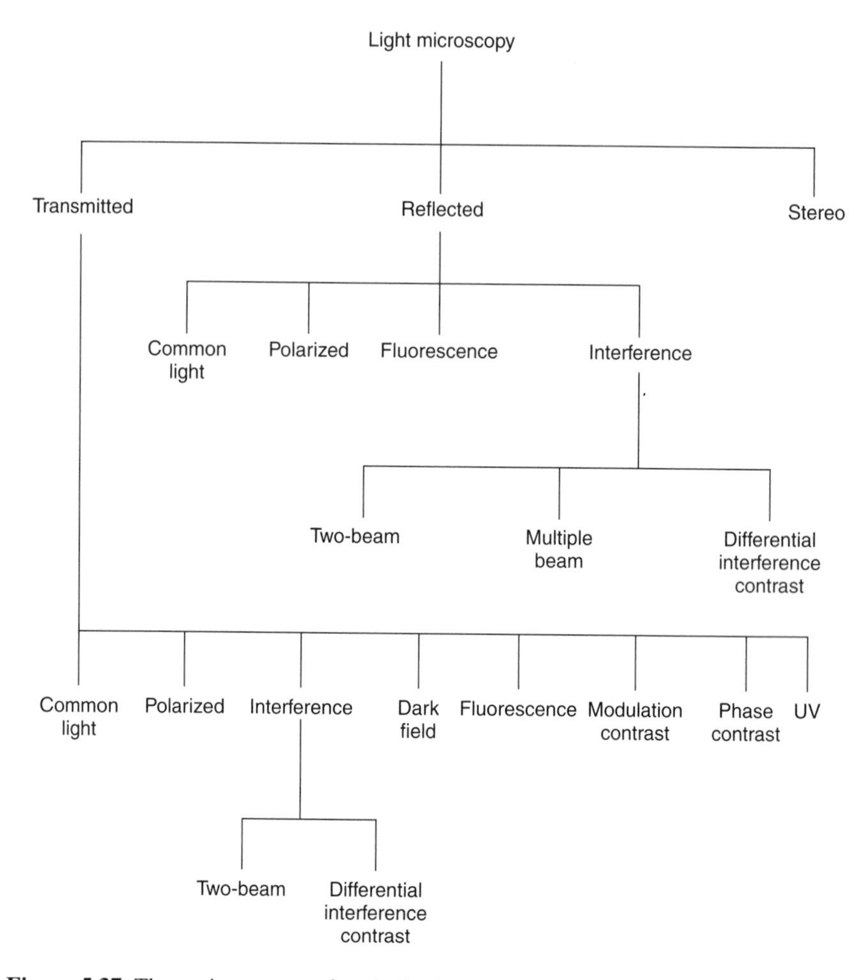

Figure 5.37 The various types of optical microscopy techniques used to examine polymers. From Sandler, S. R., Karo, W., Bonesteel, J. and Pearce, E. M., *Polymer Synthesis and Characterization: A Laboratory Manual*, © Academic Press,1998. Reproduced by permission of Academic Press.

Some polymer samples, such as powders, require minimum preparation for optical microscopy. There are various standard approaches to the preparation of bulk specimens. Many polymers can be *microtomed* (or thin-sectioned). This is best carried out at a temperature just below the T_g. Samples such as fibres or thin films can be sectioned by embedding in epoxy or acrylic resin. Thin polymers films can be produced for transmitted-light studies by pressing at an elevated temperature or by solvent-casting on to a microscope slide from a dilute (e.g. 1%) solution.

When a crystalline polymer is heated, the birefringence disappears as the crystallites disappear into the polymer melt. Changes in birefringence can be observed with a hot-stage microscope by using crossed polarizers. The point of disappearance of the last trace of birefringence can be used to determine the crystalline melting point. The rate of crystal growth of polymers may also be studied by using a hot-stage microscope, and equation (5.6) can then be applied to determine the activation energy for the process.

SAQ 5.12

A small pellet of polypropylene (PP) that had been heated on a hot-plate at 250°C for 2 min was transferred immediately to a hot-stage microscope. The diameter of one of the spherulite produced was then measured as a function of time at temperatures of 135 and 140°C, and the results obtained are listed below in Table 5.7. By plotting suitable graphs, determine the growth rates for this spherulite at both temperatures. Use the results obtained to estimate the activation energy for the diffusion of polymer chains and the experimental constant C for the process. The T_m of the PP used is 165°C.

Table 5.7 Spherulite growth data obtained for a thermally treated sample of polypropylene (SAQ 5.12)

Diameter (μm)	Time (s)	
	135°C	140°C
30	9	84
40	51	172
50	90	253
60	123	341
70	161	391
80	184	490

5.14 Transmission Electron Microscopy

Transmission electron microscopy (TEM) is a technique analogous to optical microscopy, but in this case utilizes an electron beam and electrostatic and/or electromagnetic lenses rather than a light beam and glass lenses [20]. The resolving power of TEM approaches 0.1 nm. By adjusting the power of an intermediate lens, it is possible to obtain both the image of a sample, as well as its diffraction pattern. In addition, as with optical microscopy, bright-field and dark-field imaging can be used to change the contrast.

Samples need to be in the form of very thin films – no more than 200 Å in thickness – so that the electron beam can readily pass through the sample. Ultramicrotomes may be used to cut polymer samples to the appropriate thicknesses.

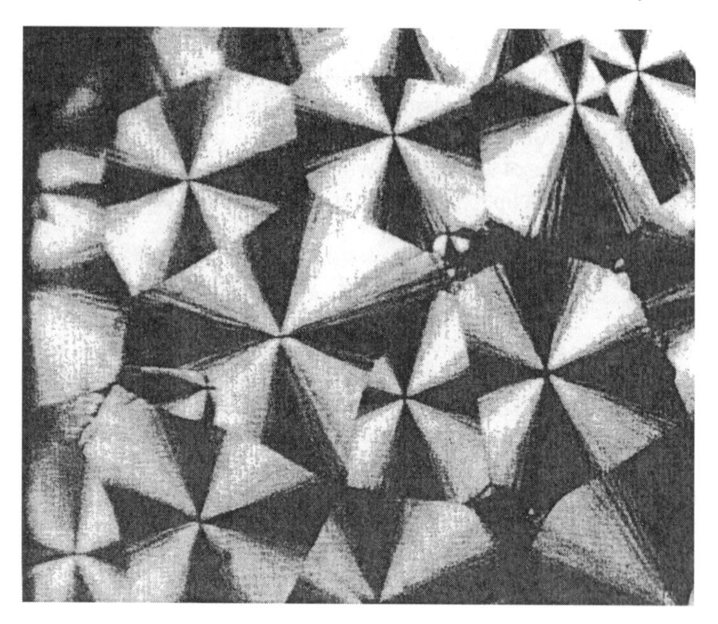

Figure 5.38 Transmission electron micrograph of polyethylene.

The electron contrast in TEM may be enhanced by the use of staining techniques. This approach involves incorporating electron-dense, heavy metals into particular sites via a chemical reaction or by absorption. For example, osmium tetroxide (OsO_4) can be reacted with polymers containing C=C bonds, with metallic osmium being deposited to provide the image contrast.

TEM is a very effective technique for the study of polymer morphology. Figure 5.38, which shows an electron micrograph of polyethylene, illustrates how TEM can be used to detail the crystalline regions of a polymer. The spherulitic structure of this sample is clearly defined. Linear boundaries form between the adjacent spherulites, with a so-called 'Maltese cross' appearing within each of them.

5.15 X-Ray Diffraction

Diffraction occurs when a wave encounters a series of regularly spaced obstacles which are capable of scattering the wave and have spacings that are comparable in magnitude to the wavelength of the radiation. X-rays have high energies and wavelengths which are of the order of the atomic spacings for solids. When a beam of X-rays encounters a solid, a portion of the beam is scattered in all directions by the electrons associated with each atom which lie within the path of the beam.

X-ray diffraction is carried out by using a *diffractometer*, illustrated schematically in Figure 5.39. The sample holder is able to be rotated. Monochromatic X-radiation is produced by bombarding a metal target with a beam of high-voltage electrons. The intensities of the diffracted beams are detected by a counter which is mounted on a movable carriage which can also be rotated. Its angular position is measured in terms of 2θ. As the counter moves at a constant angular velocity, a recorder plots the diffracted beam intensity as a function of 2θ (the *diffraction angle*).

There are two main X-ray diffraction methods commonly used to study the structures of polymers, namely *wide-angle X-ray scattering* (WAXS) and *small-angle X-ray scattering* (SAXS) [16, 22]. WAXS uses angles from 5 to 120°, and this method is useful for obtaining information about semicrystalline polymers with a range of interatomic distances from 1 to 50 Å. The size of crystals can be determined and from measurements of the relative intensities of the diffraction peaks in the crystalline region and the diffusion halo from the amorphous region, the crystallinity of a polymer can be determined. SAXS uses angles from 1 to 5° and can be employed to obtain information about polymer structures with larger interatomic distances, i.e. in the range 50–700 Å. Thus, SAXS can be used to investigate lamella structures. A schematic representation of how WAXS and SAXS diffraction patterns are produced is shown in Figure 5.40.

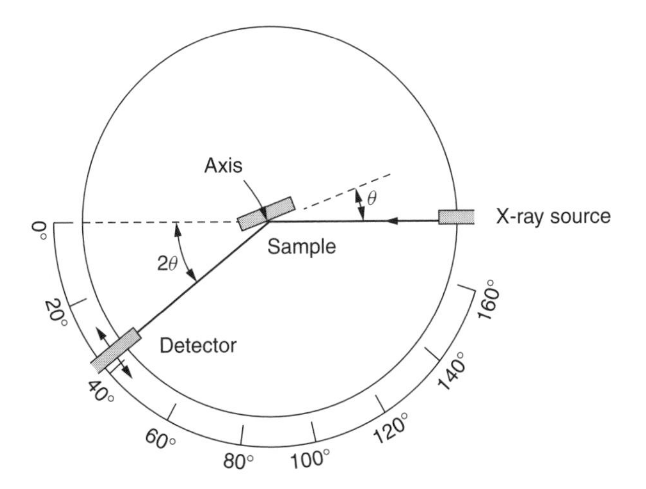

Figure 5.39 Schematic of a typical diffractometer used in X-ray diffraction studies.

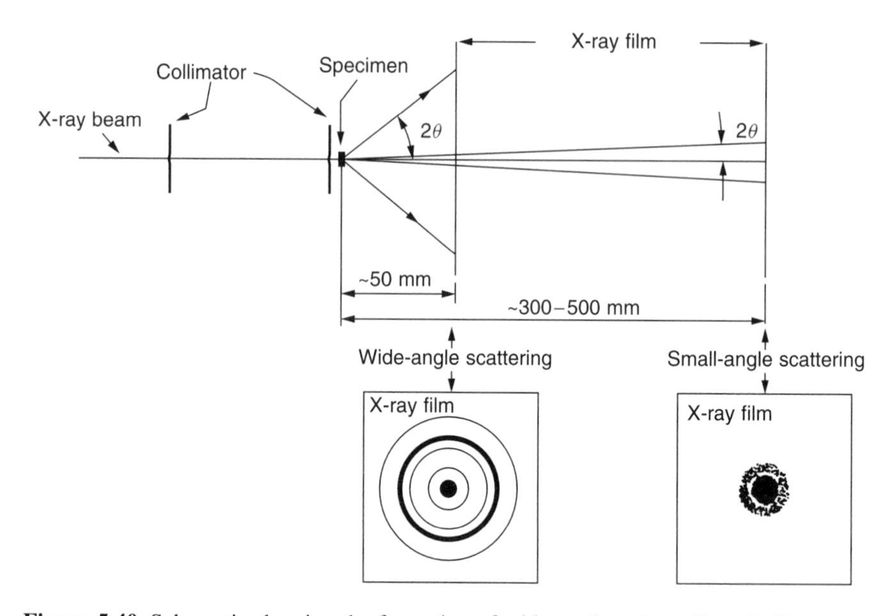

Figure 5.40 Schematic showing the formation of wide-angle and small-angle X-ray scattering diffraction patterns. From Sandler, S. R., Karo, W., Bonesteel, J. and Pearce, E. M., *Polymer Synthesis and Characterization: A Laboratory Manual*, © Academic Press, 1998. Reproduced by permission of Academic Press.

The polymer crystallites which are present in a powdered or unoriented sample diffract X-ray beams from parallel planes for incident angles (θ) determined by using the *Bragg equation*, as follows:

$$n\lambda = 2d \sin\theta \tag{5.31}$$

where λ is the wavelength of the radiation, d is the distance between the parallel planes in the crystallites and n is an integer indicating the order of diffraction (illustrated in Figure 5.41). This order parameter results from the fact that the constructive interference gives rise to a number of possible envelopes. The one that moves straight ahead in the same direction as the incident radiation is the zero-order diffracted beam, while the ones on either side are the first-order beam, second-order beam, and so on.

The reinforced waves reflected by the crystallites produce diffraction rings (or haloes). These rings appear sharply defined if the material is highly crystalline, but appear diffuse with increased amorphous content. This is illustrated in Figure 5.42, which shows the diffraction patterns of semicrystalline and amorphous poly(ether ether ketone) (PEEK). It is not often practical to produce purely crystalline or purely amorphous polymers. However, single crystals of certain polymers may be grown from solution.

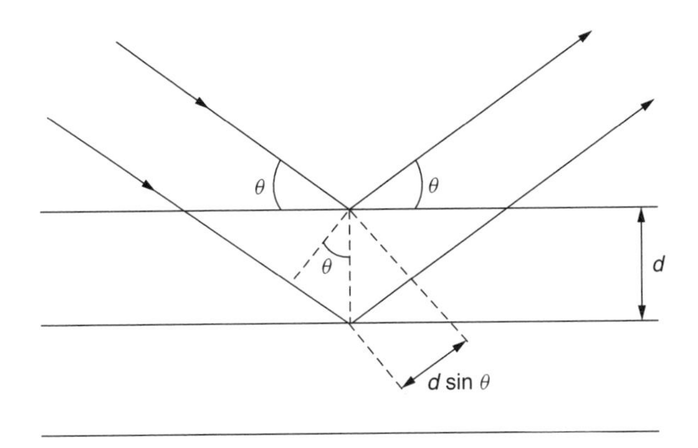

Figure 5.41 The principle of X-ray diffraction (the Bragg equation).

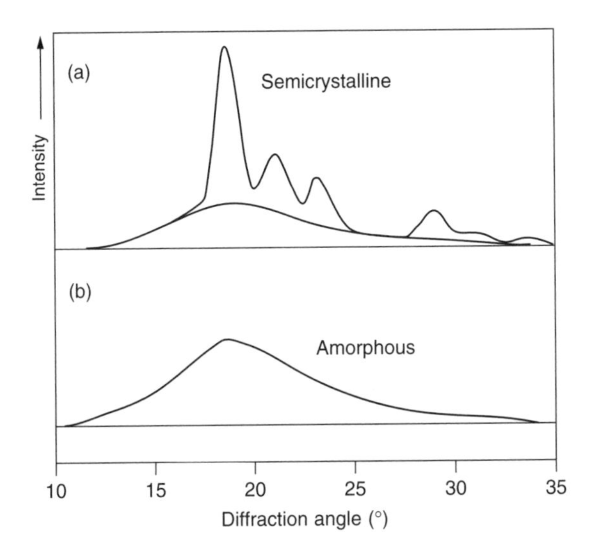

Figure 5.42 X-ray diffraction curves of (a) crystalline and (b) amorphous samples of poly(ether ether ketone).

For oriented polymers, the crystallites are aligned and the X-ray diffraction pattern is improved. This is illustrated by Figure 5.43, which shows the X-ray diffraction patterns obtained for unoriented and oriented samples of poly-oxymethylene (POM). For certain stereoregular or symmetrical polymers, the extent of ordering may be sufficient to allow structural analysis of the polymer from the X-ray data. However, in order to determine the unit cell dimensions, it is necessary to assign the crystallographic indices to the spots on the X-ray pattern, and this is particularly difficult for semicrystalline material.

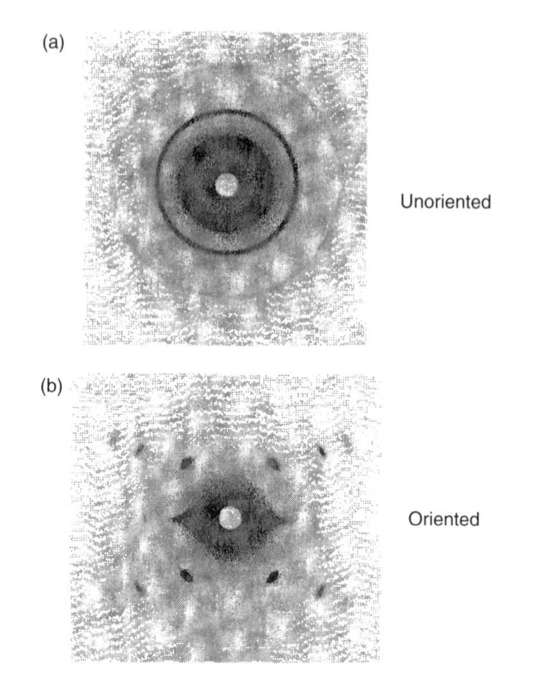

Figure 5.43 X-ray diffraction patterns of (a) unoriented and (b) oriented samples of polyoxymethylene. From F. W. Billmeyer, *Textbook of Polymer Science*, John Wiley & Sons, Inc (1984). Reproduced by permission of John Wiley & Sons, Inc.

Much X-ray work has been carried out on fibres. In certain cases, the fibres can have a well-defined periodicity parallel to the fibre axis, with more random organization in the other directions. If a set of scattering centres is considered with a repeat distance c along the chain axis, the Bragg equation then applies. The pattern on the film will consist of a set of layered lines with a spacing x, given by the following:

$$\tan\theta = x/R \tag{5.32}$$

where R is the distance between the sample and the film.

SAQ 5.13

An X-ray diffraction pattern was obtained for an oriented polymer fibre by using CuKα radiation (1.54 Å). This consisted of a series of spots arranged along a horizontal line containing a central spot and first-order lines above and below the central line. The first-order lines appear at 22.9 mm above and below the central spot. The specimen-to-film distance was 30 mm. Determine the spacing of the chain repeat unit of the polymer.

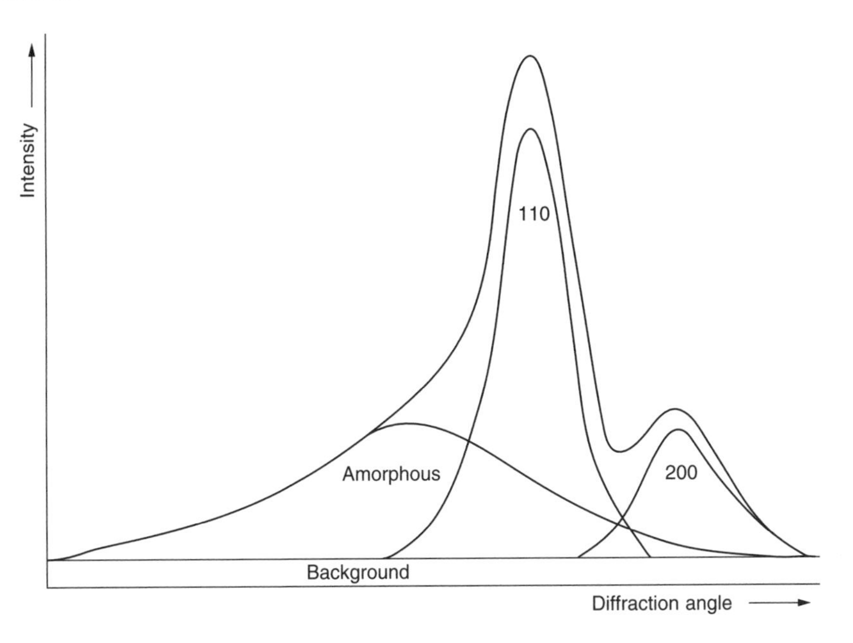

Figure 5.44 X-ray scattering curves obtained for polyethylene, illustrating the contributions from the amorphous and crystalline regions.

The crystallinity of a polymer can be determined from X-ray diffraction patterns by plotting the density of the scattered beam versus θ. If this approach is taken for a purely amorphous sample and a highly crystalline sample, the crystallinity of a sample of unknown crystallinity can be estimated. Figure 5.44 illustrates how a X-ray scattering curve for polyethylene can be resolved into the various contributions from the amorphous and crystalline regions. The relatively sharp peaks are due to scattering from the crystalline regions, while the broad underlying peak is due to amorphous scattering. The mass fraction of crystals (x_c) can be approximated by using the following equation:

$$x_c = A_c/(A_a + A_c) \tag{5.33}$$

where A_a is the area of the amorphous peak and A_c is the area remaining under the crystalline peaks.

5.16 Neutron Scattering

As with X-ray scattering, neutron scattering may be employed to provide size information about polymers on a much smaller scale than is possible by using light scattering [23, 24]. The wavelengths of the neutrons used in this technique are typically of the order of 0.1–2.0 nm. However, in order to obtain reliable data, very small scattering angles ($\theta < 2^\circ$) are required. This also means that the detector must be positioned a long distance away from the sample (typically 1–40 m). The same basic theory to that used in light scattering can be employed. The major practical restriction for the *small-angle neutron scattering* (SANS) technique is the requirement of a source of neutrons direct from a nuclear reactor.

SANS has been successfully applied to the study of solid polymers, and particularly to studies of chain dimensions [23]. Polymer morphology has been commonly studied by using this technique, while the size of dispersed phases in polymers may also be determined. Such measurements are made by incorporating a small amount of deuterated polymer into the polymer sample. This provides the necessary contrast for SANS, as there is a large difference between the neutron scattering lengths of ^{1}H and ^{2}H atoms. The dimensions of individual blocks in copolymers can also be determined by using such an approach.

Summary

A variety of different structures in polymers gives rise to a wide range of physical properties. Different isomers and conformations for polymeric materials were therefore considered in this chapter. The presence of crystalline regions in thermoplastics plays an integral role in their physical properties, and so the forms of crystallinity in polymers were also discussed. Polymer blends are used to modify and improve the properties of the individual constituent materials and were described. The properties of polymers are temperature-dependent and thus the thermal behaviour of polymeric systems was outlined in detail. A broad range of techniques are used to investigate the various structures of polymers, with the following being covered in this chapter: dilatometry, infrared spectroscopy, Raman spectroscopy, NMR spectroscopy, differential scanning calorimetry, thermal mechanical analysis, dynamic mechanical analysis, optical microscopy, transmission electron microscopy, X-ray diffraction and neutron scattering.

References

1. Bassett, D. C., *Principles of Polymer Morphology*, Cambridge University Press, Cambridge, UK, 1981.
2. Percec, V. and Tomazos, D., 'Molecular Engineering of Liquid Crystalline Polymers', in *Comprehensive Polymer Science*, 1st suppl. Aggarwal, S. L. and Russo, S. (Eds), Pergamon Press, Oxford, UK, 1992, pp. 299–385.
3. White, J. L. and Cakmak, M., 'Orientation', in *Encyclopedia of Polymer Science and Engineering*, Vol. 10, Mark, H. F. (Ed.), Wiley, New York, 1987, pp. 595–618.
4. Walsh, D. J., 'Polymer Blends', in *Comprehensive Polymer Science*, Vol. 2, Booth, C. and Price, C. (Eds), Pergamon Press, Oxford, UK, 1989, pp. 135–154.
5. Thompson, E. V., 'Thermal Properties', in *Encyclopedia of Polymer Science and Engineering*, Vol. 16, Mark, H. F. (Ed.), Wiley, New York, 1987, pp. 711–747.
6. Runt, J. P., 'Crystallinity Determination', in *Encyclopedia of Polymer Science and Engineering*, Vol. 4, Mark, H. F. (Ed.), Wiley, New York, 1987, pp. 482–519.

7. Koenig, J. L., *Spectroscopy of Polymers*, Elsevier, Amsterdam, The Netherlands, 1999.
8. Bower, D. I. and Maddams, W. F., *The Vibrational Spectroscopy of Polymers*, Cambridge University Press, Cambridge, UK, 1989.
9. Schroder, E., Muller, G. and Arndt, K. F., *Polymer Characterization*, Hanser, Munich, 1989.
10. Painter, P. C. and Coleman, M. M., *Fundamentals of Polymer Science: An Introductory Text*, Technomic Publishing, Lancaster, PA, 1997.
11. Moskala, E. J., Varnell, D. F. and Coleman, M. M., *Polymer*, **26**, 228–234 (1985).
12. Hendra, P., Jones, C. and Warnes, G., *Fourier Transform Raman Spectroscopy: Instrumentation and Chemical Applications*, Ellis Horwood, New York, 1991.
13. Jackson, K. D. O., Loadman, M. J. R., Jones, C. H. and Ellis, G., *Spectrochim. Acta, A*, **46**, 217–226 (1990).
14. Briscoe, B. J., Stuart, B. H., Thomas, P. S. and Williams, D. R., *Spectrochim. Acta, A*, **47**, 1299–1303 (1991).
15. Young, R. J. and Lovell, P. A., *Introduction to Polymers*, Chapman and Hall, London, 1991.
16. Sandler, S. R., Karo, W., Bonesteel, J. and Pearce, E. M., *Polymer Synthesis and Characterization: A Laboratory Manual*, Academic Press, San Diego, CA, 1998.
17. Bovey, F. A., *High Resolution NMR of Macromolecules*, Academic Press, New York, 1972.
18. Haines, P. J., *Thermal Methods of Analysis: Principles, Applications and Problems*, Blackie, London, 1995.
19. Brandrup, J., Immergut, E. H. and Grulke, E. A. (Eds), *Polymer Handbook*, Wiley, New York, 1999.
20. Vaughan, A. S., 'Polymer Microscopy', in *Polymer Characterization*, Hunt, B. J. and James, M. I. (Eds), Blackie, London, 1993, pp. 297–332.
21. Sawyer, L. C. and Grubb, D. T., *Polymer Microscopy*, Chapman and Hall, London, 1987.
22. Atkins, E. D. T., 'Crystal Structure by X-ray Diffraction', in *Comprehensive Polymer Science*, Vol. 1, Booth, C. and Price, C. (Eds), Pergamon Press, Oxford, UK, 1989, pp. 613–650.
23. Higgins, J. S. and Benoit, H. C., *Polymers and Neutron Scattering*, Clarendon Press, Oxford, UK, 1996.
24. Richards, R. W., 'Small Angle Neutron Scattering and Neutron Reflectometry', in *Polymer Characterization*, Hunt, B. J. and James, M. I. (Eds), Blackie, London, 1993, pp. 222–260.

Chapter 6

Surface Properties

Learning Objectives

- To apply infrared reflectance techniques, Raman spectroscopy, photoelectron spectroscopy, secondary-ion mass spectrometry, inverse gas chromatography, scanning electron microscopy, surface tension techniques, atomic force microscopy and tribological analysis to the study of polymer surfaces.

6.1 Introduction

The surface properties of polymers are of great importance in many of the applications in which these materials are employed, including adhesives, lacquers, metallization and composites. In fact, unsuitable surfaces properties may negate the otherwise advantageous bulk properties of a particular polymer. For instance, poor adhesive properties will render a polymer unsuitable for use as a surface coating. Surface modification by chemical means is also a common pretreatment in, for instance, biomedical polymers, and an understanding of such treatments is therefore important.

The techniques used for studying the surfaces of polymers have been available for a number of years. Infrared spectroscopic methods, such as attenuated total reflectance, diffuse reflectance, specular reflectance and photoacoustic spectroscopy, plus Raman spectroscopy, allow the surface properties in the order of microns to be characterized. More recently, with the advent of improved vacuum techniques, more sensitive surface characterization of the order of angstroms and nanometres has been possible by using techniques such as X-ray photoelectron spectroscopy and secondary-ion mass spectrometry. As it is believed that polymer surface behaviour is determined by a surface layer of less than 10 nm in thickness, it is clearly important to differentiate the properties of this thin layer from the bulk properties. In addition, techniques such as inverse gas chromatography, scanning electron microscopy, surface tension measurements, atomic force microscopy and tribological analysis provide information about a range of surface properties of polymers important to their specific applications.

6.2 Infrared Spectroscopy

Reflectance techniques can be used for samples which are difficult to analyse by conventional transmittance methods [1]. Reflectance methods can be divided into two categories. Internal reflectance measurements can be made by using an *attenuated total reflectance* cell in contact with the sample. There are also external reflectance measurements which involve an infrared beam reflected directly from the sample surface. In external reflectance, incident radiation is focused on to the sample and two forms of reflectance can occur, i.e. *specular* and *diffuse*.

6.2.1 Attenuated Total Reflectance Spectroscopy

Attenuated total reflectance (ATR) spectroscopy utilizes the phenomenon of total internal reflection (Figure 6.1). A beam of radiation entering a crystal will undergo total internal reflection when the angle of incidence at the interface between the sample and crystal is greater than the *critical angle*, where the angle is a function of the refractive indices of the two surfaces. The beam penetrates a fraction of a wavelength beyond the reflecting surface and when a material which selectively absorbs radiation is in close contact with this surface, the beam loses energy at the wavelength where the material absorbs. The resultant attenuated radiation is measured and plotted as a function of wavelength by the spectrometer and gives rise to the absorption spectral characteristics of the sample.

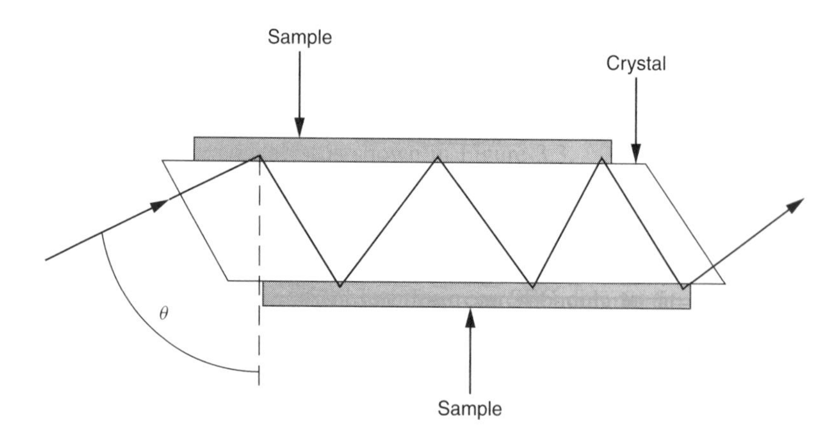

Figure 6.1 Schematic of an attenuated total reflectance cell.

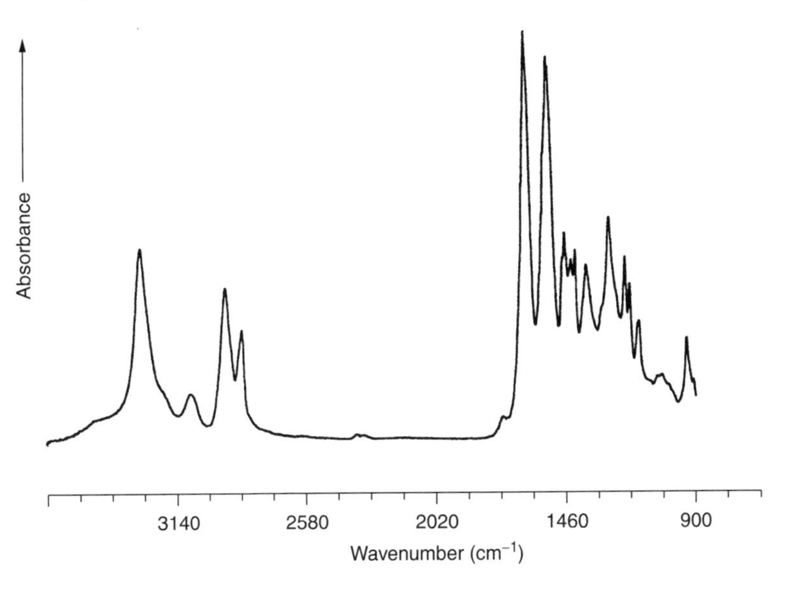

Figure 6.2 An ATR spectrum of a nylon 6 film.

The depth of penetration in ATR is a function of the wavelength (λ), the refractive index of the crystal and the angle of incident radiation (θ). The depth of penetration (d_p) for a non-absorbing medium is given by the following formula:

$$d_p = (\lambda/n_1)/\{2\pi[\sin\theta - (n_1/n_2)^2]^{1/2}\} \qquad (6.1)$$

where n_1 is the refractive index of the sample and n_2 is the refractive index of the ATR crystal. Thus, when compared to a conventional infrared absorbance spectrum, the bands at lower wavenumbers appear more intense. The ATR spectrum of a nylon 6 film is shown in Figure 6.2. The crystals used in ATR cells are made from materials that are of high refractive index, such as zinc selenide, germanium and thallium iodide (KRS-5). It has been shown that the sampling depth for polymeric materials is about three times the d_p [2]. The sampling depths for a polymer that has a refractive index of 1.5, examined by using a 45° ZnSe ATR element, have been calculated at various frequencies and are listed in Table 6.1. The sampling depth in this case is thus typically 2–6 μm.

SAQ 6.1

The infrared spectrum of the nylon 6 film (refractive index of 1.5) shown in Figure 6.2 was produced by using an ATR cell made of KRS-5 (refractive index of 2.4). If the incident radiation enters the cell crystal at an angle of 60°, what is the depth of penetration into the sample surface at (a) 1000, (b) 1500, and (c) 3000 cm^{-1}?

Table 6.1 The sampling depths for a polymer with a refractive index of 1.5, obtained by using a 45° ZnSe ATR element

Wavenumber (cm^{-1})	Sampling depth (μm)
900	6.3
1600	3.5
3000	1.9

6.2.2 Specular Reflectance Spectroscopy

Specular reflectance occurs when the reflected angle of incident radiation equals the angle of incidence. The amount of light reflected depends on the angle of incidence, the refractive index, the surface roughness and the absorption properties of the sample. For most materials, the reflected energy is only 5–10%, but in regions of strong absorptions the reflected intensity is greater. The resulting data appear different from normal transmission spectra, as derivative-like bands result from the superposition of the normal extinction coefficient spectrum with the refractive index dispersion (based upon Fresnel's relationships). However, the reflectance spectrum can be corrected by using a *Kramers–Kronig transformation* (or K–K transformation). The corrected spectrum then appears like the familiar transmission spectrum.

6.2.3 Diffuse Reflectance Spectroscopy

In external reflectance, the energy that penetrates one or more particles is reflected in all directions. This component is described by the term *diffuse reflectance*. In this technique, commonly called DRIFT, a powdered sample is mixed with KBr powder. The DRIFT cell reflects radiation to the powder and collects the energy reflected back over a large angle. Diffusely scattered light can be collected directly from a sample or, alternatively, by using an abrasive sampling pad. DRIFT is particularly useful for sampling powders or fibres. Figure 6.3 illustrates diffuse reflectance from a surface.

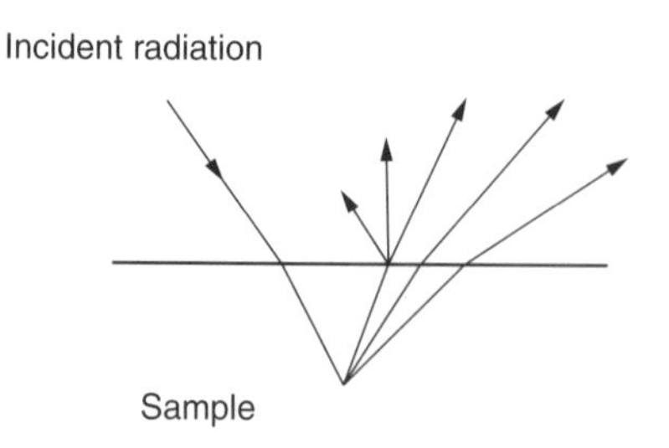

Figure 6.3 Illustration of diffuse reflectance from a surface.

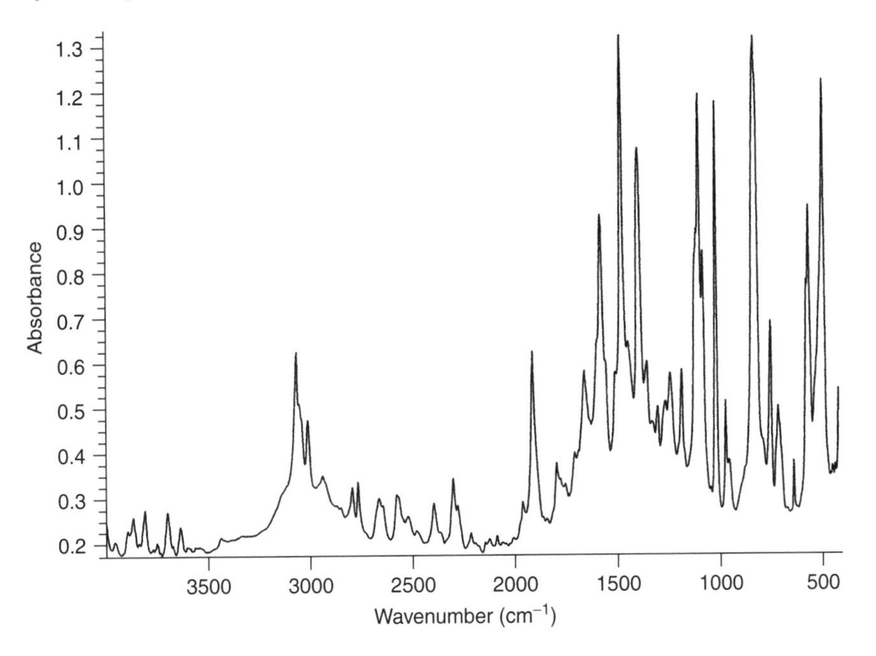

Figure 6.4 Diffuse reflectance spectrum of a sample of poly(phenylene sulfide) mixed with KBr powder.

Kubelka and Munk developed a theory describing the diffuse reflectance process for powdered samples which relates the sample concentration to the scattered radiation intensity. The *Kubelka-Munk* equation is given by the following:

$$(1 - R_{\infty})^2/2R_{\infty} = c/k \tag{6.2}$$

where R_{∞} is the absolute reflectance of the layer, c is the concentration and k is the molar absorption coefficient. Figure 6.4 shows the diffuse reflectance spectrum of a sample of poly(phenylene sulfide) (PPS) mixed with KBr powder.

SAQ 6.2

Diffuse reflectance spectroscopy was used to examine thin films of poly(methyl methacrylate) (PMMA), spin-coated on to high-density polyethylene (HDPE) substrates [3]. The film thicknesses ranged from 0.003 to 2 μm. It was found that over this range a plot of the logarithm of the absorbance of the PMMA carbonyl band at 1730 cm^{-1} versus the logarithm of the PMMA layer thickness is linear. The absorbance values of the carbonyl bands of the 0.003 and 2 μm films are 6.17×10^{-3} and 0.380, respectively. Determine the thickness of a PMMA film showing a carbonyl band of absorbance 1.34.

6.2.4 Photoacoustic Spectroscopy

Photoacoustic spectroscopy (PAS) is based on the transfer of modulated infrared radiation to a mechanical vibration. Gaseous, liquid or solid samples can be measured by using PAS and the technique is particularly useful for highly absorbing samples such as rubber and polyacetylene. When the modulated infrared radiation is absorbed by a sample, the substance heats and cools in response to the radiation reaching the material. This heating and cooling is converted into a pressure wave that can be detected by a microphone.

The resulting spectrum differs from either a transmittance or a reflectance spectrum since the technique detects non-radiative transitions in the sample. PAS is useful because the detected signal is proportional to the sample concentration and can be used with very black or highly absorbing samples. This spectroscopic technique probes mainly material on the surface or several microns below the surface of the sample and so is very useful for surface studies. In addition, by changing the modulation range, a depth profile of the sample may be obtained and so sub-surface layers can be characterized.

6.3 Raman Spectroscopy

Raman spectroscopy is a useful technique for the study of thin surface films because of the non-destructive nature of this approach. This technique also provides an advantage over other methods, such as infrared spectroscopy, in the study of paint films. Many coating materials contain pigments or are water-based and can therefore be difficult to examine in the infrared region. Fourier-transform (FT) Raman spectroscopy has been used to study paint systems [4, 5]. For example, most oil-based paints and coatings are based on alkyd resins and the curing of such resins can be monitored by observation of the C=C stretching band. The emulsion polymerization of a butyl acrylate/methyl methacrylate/allyl methacrylate copolymer latex has been studied by using Raman spectroscopy. The acrylic and methacrylic C=C bonds react to form the polymer chains, while the allylic C=C bonds produce cross-links. As the C=C stretching is strong in Raman spectroscopy, the extent of polymerization can be determined by monitoring the disappearance of this band.

6.4 Photoelectron Spectroscopy

X-ray photoelectron spectroscopy (XPS) (or Electron Spectroscopy for Chemical Analysis (ESCA)) has provided much information regarding the surface elemental and fundamental group compositions of polymers [6–8]. The technique is based on the principle that when a sample is irradiated with soft X-rays, usually MgKα (1253.6 eV) or AlKα (1486.6 eV), photoelectrons are emitted. Depending on the

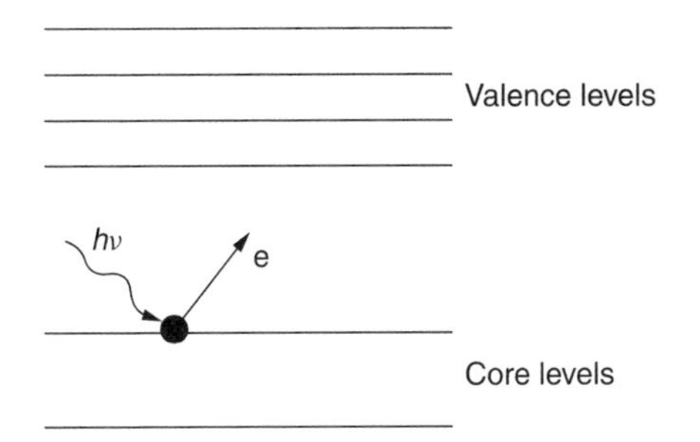

Figure 6.5 The generation of photoelectrons from core (or valance) levels in a sample after irradiation with soft X-rays (photoelectron spectroscopy).

energy of the X-ray source being used, photoelectrons can arise from the core or valence levels, as shown in Figure 6.5. On removal of the electron, a vacancy remains, which can be filled by an electron from a higher level. The energy released in filling the vacancy may result in the emission of an X-ray or may be transferred to another weakly bound electron. The latter is emitted as Auger electrons, which although element-specific, are not regarded as particularly useful in polymer characterization.

XPS experiments are conducted in a high-vacuum chamber. When the sample is irradiated, the emitted photoelectrons are collected by a lens system and focused into an energy analyser. The latter counts the number of electrons with a given kinetic energy (E_K). The binding energies (E_B) of the photoelectrons are obtained by using the Einstein equation as follows:

$$E_B = h\nu - E_K - \Phi \tag{6.3}$$

where $h\nu$ is the X-ray photon energy and Φ is the work function of the sample.

The core electron binding energies are characteristic of the atomic core levels from which the photoelectrons are emitted and thus enable the surface elemental composition to be determined. All elements except hydrogen may be detected. Carbon bound to itself and/or hydrogen only gives C 1s = 285 eV. The binding energies of oxygen O 1s fall near 533 eV, while nitrogen N 1s falls near 400 eV. An example of a photoelectron spectrum of poly(ethylene terephthalate) (PET) is shown in Figure 6.6. There are three peaks observed at 286, 534 and 990 eV, which correspond to the C 1s, O 1s and oxygen Auger lines, respectively.

The core binding energies in an atom are influenced by the local electronic environment, and consequently, an atom in a molecule can exhibit a small range of binding energies – known as *chemical shifts*. For example, oxygen induces shifts in the binding energy of carbon by approximately 1.5 eV per C–O bond, while halogens induce shifts to higher binding energies for carbon in the range 1–3 eV. Nitrogen-containing substituents cause C 1s shifts up to 2 eV, depending on the nature of the substituent. Oxygen binding energies are shifted over a

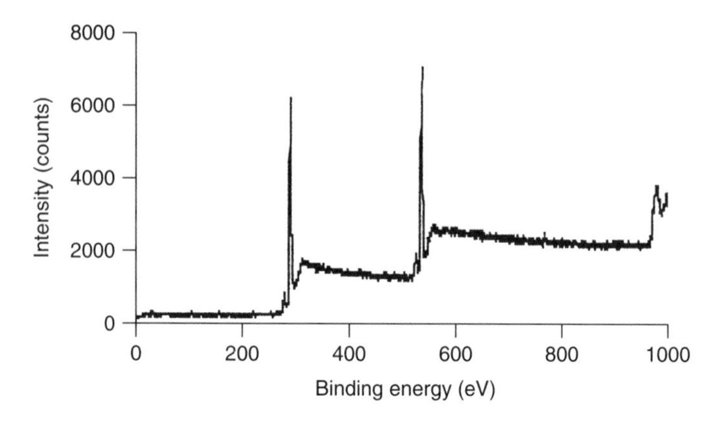

Figure 6.6 Photoelectron spectrum of poly(ethylene terephthalate).

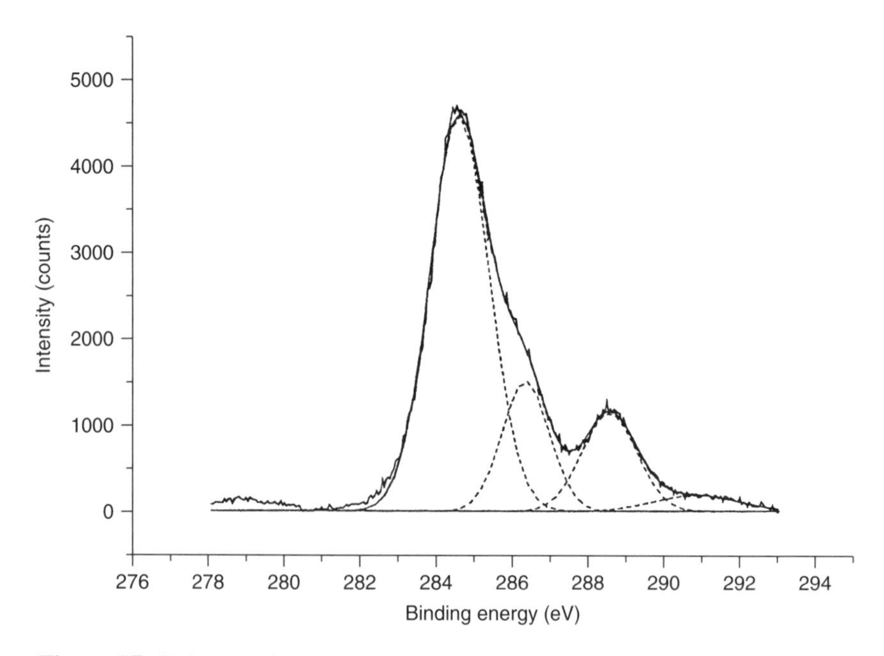

Figure 6.7 Carbon 1s photoelectron spectral profile of poly(ethylene terephthalate).

narrow range of 2 eV near 533 eV. Carboxyl and carbonate groups have a singly bound oxygen at a higher binding energy. The chemical shift data for a wide range of polymers has been collected and is available in various publications [9, 10].

The XPS spectrum shown in Figure 6.7 illustrates the peak-fitted C 1s region for poly(ethylene terephthalate) (PET). The four peaks observed indicate the presence of four different types of bonding in the material. The largest peak at 284.7 eV

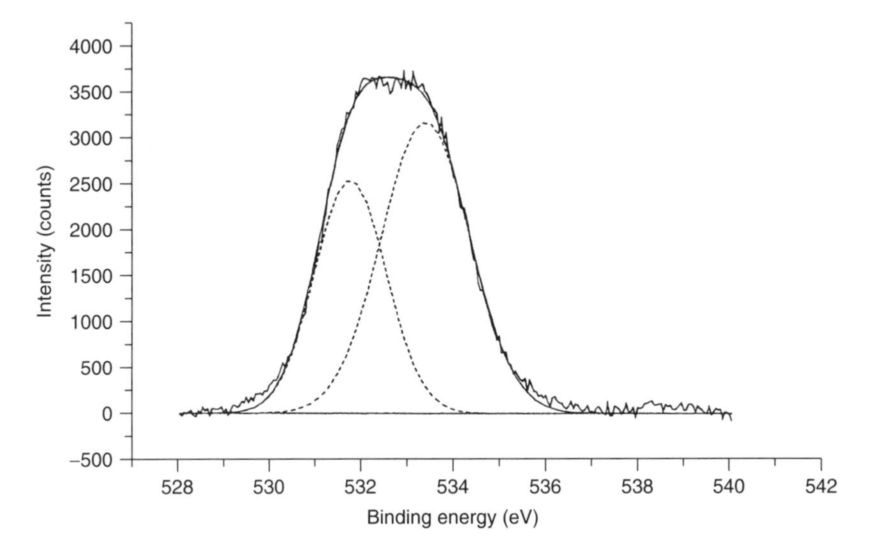

Figure 6.8 Oxygen 1s photoelectron spectral profile of poly(ethylene terephthalate) (DQ 6.1).

corresponds to C–C bonding, while the peaks at 286.5 and 288.5 eV correspond to C–O bonding and O–C=O bonding, respectively. The small broad peak at 291.1 eV is a 'shakeup satellite' which occurs as a result of the aromatic ring in the PET structure.

DQ 6.1

Figure 6.8 above shows the curve-fitted O 1s profile of the X-ray photoelectron spectrum of PET. If the peaks at 531.9 and 533.6 eV correspond, respectively, to the C=O and C–O bonds, what do the relative areas of these binding energies imply about the structure of this polymer?

Answer

The ratio of the two peaks resulting from the curve-fitted oxygen 1s profile is approximately 1:1. This agrees with the 1:1 ratio of C=O and C–O bonds in the PET structure.

SAQ 6.3

In order to produce a modified PET sample suitable for use as a biomaterial, the surface of the polymer film was reacted with an amine [11]. XPS was carried out on the treated PET film and the resulting C 1s profile is shown below in Figure 6.9. What does this spectrum imply about the treated polymer film?

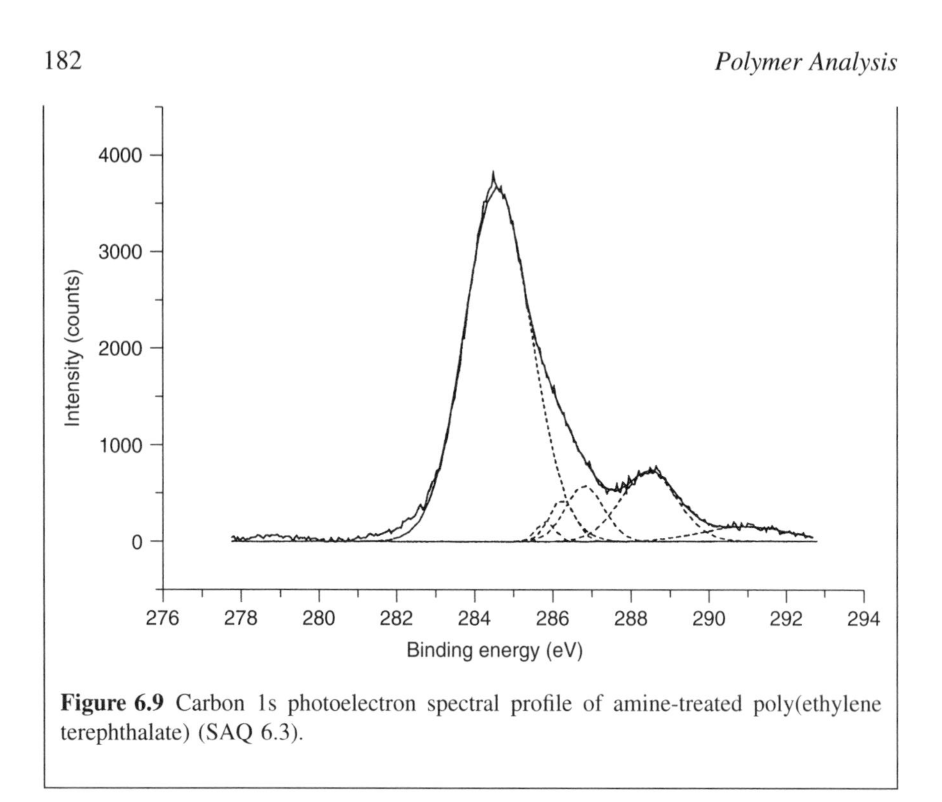

Figure 6.9 Carbon 1s photoelectron spectral profile of amine-treated poly(ethylene terephthalate) (SAQ 6.3).

The surface sensitivity of XPS arises from the limited distance that an electron with a given kinetic energy can travel (elastically) through a material. The sampling depth of XPS refers to the depth from which 95% of the signal intensity arises, and is given by the following:

$$d = 3\lambda \cos \theta \tag{6.4}$$

where d is the sampling depth, λ is the inelastic mean free path, and θ is the angle between the energy analyser and the sample surface.

6.5 Secondary-Ion Mass Spectrometry

Static *secondary-ion mass spectrometry* (SIMS) involves the bombardment of a sample contained in an ultra-high vacuum chamber by a primary ion beam [7, 9]. Positively and negatively charged atoms or molecular fragments are *sputtered* from the surface during this process. The static mode refers to the fact that the accumulated ion dose during acquisition is low enough for the surface to be essentially unchanged during the measurement. The bombardment energies vary from 2 to 30 eV, and noble gases such as Ar and Xe are commonly used. While initial SIMS instrumentation employed quadruple mass spectrometers, in recent times, time-of-flight (TOF) mass spectrometers have become more significant as they offer an unlimited mass range and better resolution. Usually, the primary ion penetration depth ranges from nanometres to multiples of tens of nanometres. The resulting spectrum shows the intensity in counts per second as a function of the mass (in daltons (DA)) .

SIMS provides a useful means of identifying the composition of polymer surfaces [12]. Generally, the secondary-ion mass spectra of homopolymers contain peaks resulting from main-chain scission and intact side-chains. This allows for different classes of polymers to be identified, but also provides for the individual polymers of a class to be differentiated. For example, SIMS can be used to distinguish different types of polymethacrylates. In the positive-ion spectra of methacrylate esters, the region above 80 Da is similar for all derivatives as it is characteristic of the methacrylate backbone. However, below 80 Da the spectrum shows fragments from the ester side-chain. For example, distinctive peaks at 15, 29, 45 and 57 Da are characteristic of the methyl, ethyl, hydroxyethyl and butyl groups in the ester side-chain, respectively. The negative-ion spectrum may also be used to distinguish between different types of polymethacrylates.

6.6 Inverse Gas Chromatography

Inverse gas chromatography (IGC) is a form of chromatography which uses a polymer as the stationary phase, and where the interaction with a solvent is measured [13]. Polymer surfaces can be studied through an understanding of the magnitude and temperature-dependence of this interaction. Conventional gas chromatographs can be used for carrying out IGC. The column can be packed with the polymer dispersed as a thin film on the surface of an inactive support, or packed directly with the polymer in the form of a fibre, film or powder. Inert gas flows through the column, and probe molecules are injected at one end of the column and detected at the other end.

Solute retention data are determined from the *specific retention volume* (V_g) by using the following relationship:

$$V_g = 273.15/T(V_R/w) = 273.15/T(t_r - t_m)/(wFJ) \tag{6.5}$$

where V_R is the retention volume at temperature T, w is the weight of the stationary phase in the column, t_r is the elution time of the interacting solute, t_m is the elution time of the non-interacting gas, F is the flow rate of the carrier gas through the column, and J is the gas compressibility factor. The dead volume may be measured by injecting a non-interacting solute, such as methane, into the column and measuring the corresponding retention time.

A *retention diagram* is obtained by plotting the logarithm of V_g as a function of $1/T$. The slope of this plot is related to the enthalpy of the solution in the stationary phase or the adsorption on the solid surface, as follows:

$$\mathrm{d}\ln V_g/\mathrm{d}(1/T) = -\Delta H/R \tag{6.6}$$

where ΔH is the enthalpy and R is the universal (molar gas) constant. Deviations from this slope can be used to obtain information regarding various thermal transitions.

The adsorption isotherm for a polymer, which describes the variation in the amount of solute adsorbed with its pressure in the bulk phase at constant temperature, can be evaluated. The heat of adsorption and the surface area of the polymer stationary phase can also be obtained from experimental adsorption isotherms.

Figure 6.10 shows an IGC retention diagram for PVC-coated 'Kevlar' fibres exposed to dodecane vapours [14]. This figure indicates that an inflection is observed at a critical temperature – which is coincident with the T_g (84°C). The notable differences in the slopes below and above the T_g can be attributed to the fact that below the T_g, the primary mechanism for vapour retention is surface adsorption due to the relatively rigid nature of the polymer. However, once the temperature exceeds the glass transition temperature, the change to a softer nature allows for the bulk sorption of vapour to occur in addition to surface adsorption.

SAQ 6.4

Using Figure 6.10, determine the enthalpies of adsorption of dodecane on to the PVC-coated fibre above and below the T_g.

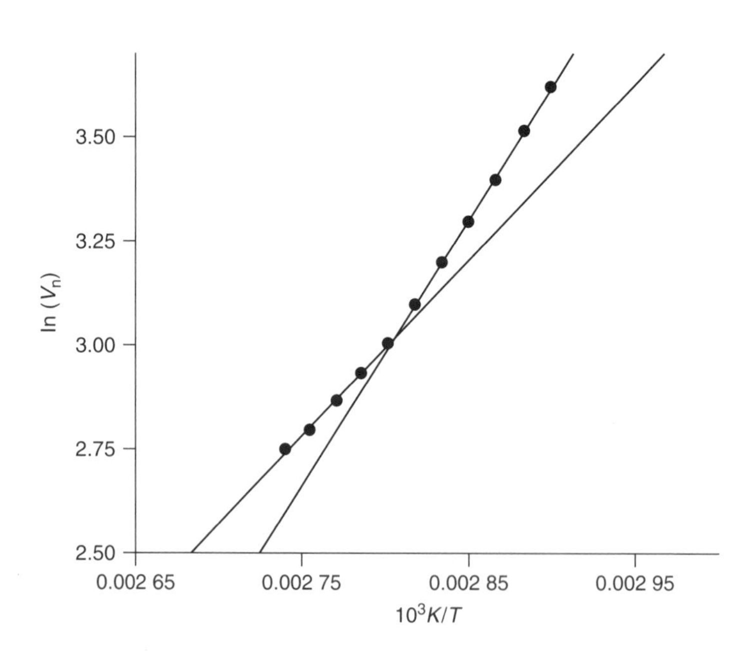

Figure 6.10 An inverse gas chromatography retention diagram obtained for PVC-coated 'Kevlar' fibres exposed to dodecane vapours.

6.7 Scanning Electron Microscopy

An image of the surface of a polymer can be produced by using *scanning electron microscopy* (SEM) [15]. In this technique, a fine electron beam (5–10 nm in diameter) is scanned across the sample surface in synchronization with a beam from a cathode-ray tube. The scattered electrons produced can then result in a signal which modulates this beam. This produces an image with a depth-of-field which is usually 300–600 times better than that of an optical microscope, and also enables a three-dimensional image to be obtained. Most scanning electron microscopes have magnification ranges from ×20 to ×100 000. As polymers tend not to be good conductors, they need to be coated with a thin layer of a conducting material such as gold. However, in recent years a new *environmental scanning electron microscopy* (ESEM) technique has emerged which enables samples to be investigated without the need to apply any conducting coating.

SEM has been used for a broad range of polymer studies and applications, including surface roughness, adhesive failures, fractured surfaces, networks, and phase boundaries in blends. Figure 6.11 presents an environmental scanning electron micrograph of an amine-treated biaxially oriented PET film. Rows of lamellae can be observed, and the amorphous regions between the rows and between the individual lamellae themselves can also be identified in the micrograph.

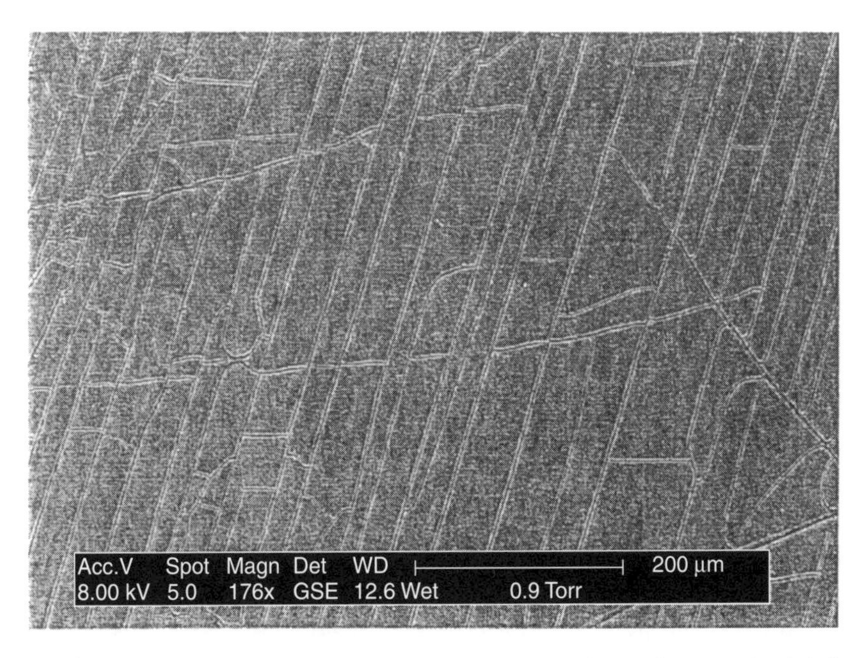

Figure 6.11 Environmental scanning electron micrograph of an amine-treated poly(ethylene terephthalate) film.

6.8 Surface Tension

Surface tension results from unbalanced cohesive forces of molecules at the surface of a material, e.g. a polymer. Most commonly, the surface tension of polymers is determined by using *contact angle* measurements [16]. This type of experiment involves measuring the contact angle (θ) of a series of liquids of known surface tension (γ_L) on plane solid surfaces. As illustrated in Figure 6.12, different values of θ can lead to different types of behaviour, as follows:

(a) $\theta = 0$ – the liquid wets the solid and spreads over the surface;
(b) $0 < \theta < \pi/2$ – the liquid spreads on the surface over a limited area;
(c) $\theta > \pi/2$ – the liquid does not wet the surface and tends to shrink away from the solid.

A plot of $\cos\theta$ versus γ_L provides the critical surface tension of the polymer (γ_c). Table 6.2 presents the critical surface tension values for a number of polymers at 20°C [16].

The rate of spreading of a liquid on a polymer surface may be determined by using the following equation:

$$dA/dt = k\gamma_L(\cos\theta_s - \cos\theta_d) \tag{6.7}$$

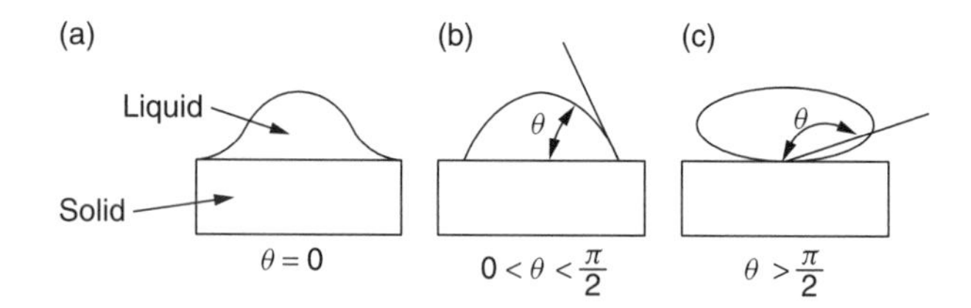

Figure 6.12 Different contact angles leading to different types of behaviours for liquids on solid surfaces.

Table 6.2 Critical surface tension values of a number of common polymers

Polymer	γ_c (mN m^{-1})
PTFE	19
LDPE	31
PP	31
PMMA	39
PVC	39
PS	43
PET	43
Nylon 6,6	46

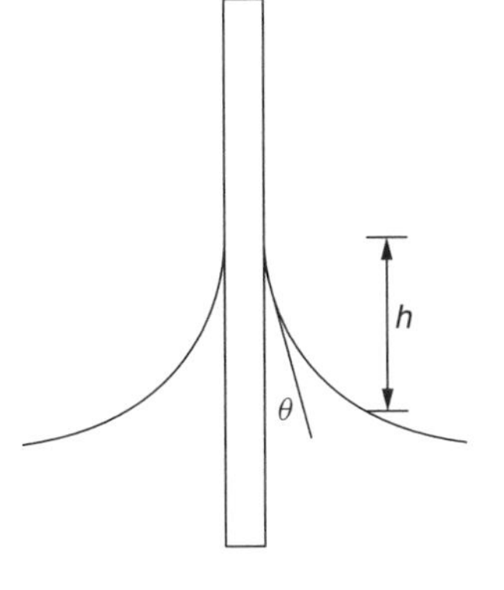

Figure 6.13 Schematic of the slide technique used for measuring contact angles.

where dA/dt is the rate of increase in the liquid–solid contact area, k is the rate constant, γ_L is the surface tension of the liquid, θ_d is the dynamic contact angle, and θ_s is the static contact angle (or equilibrium contact angle) measured over a prolonged contact time.

A common method for the measurement of the contact angle is the *slide technique*. In this method, a polymer plate is immersed in the liquid, as illustrated in Figure 6.13. The meniscus of a partially immersed plate rises to a specific height (h) if θ is finite and can be measured with a travelling microscope. The contact angle may then be calculated by using the following expression:

$$\sin\theta = 1 - (h/a)^2 \tag{6.8}$$

where a is a capillary constant.

6.9 Atomic Force Microscopy

Atomic force microscopy (AFM) is a type of scanning probe microscopy which allows the topography of a polymer surface to be mapped [17]. Polymer surfaces can be examined to resolutions in the angstrom range by using this technique. The atomic force microscope consists of a very fine tipped probe which is positioned several angstroms above the surface of the (polymer) material. The tunnelling current between the tip of the probe and the surface is then measured. The probe, usually made of silicon nitride, is attached to a cantilever with a reflective surface. A piezo-electric support is used to mount the sample and moves in response to surface changes sensed by the probe. The resulting deflections are monitored by the reflected laser beam. Measurements can be made in two different modes, i.e. in *contact mode*, without oscillation of the cantilever, and *tapping mode*, with oscillation of the cantilever. A three-dimensional representation of a polymer surface obtained by using atomic force microscopy is given in Figure 6.14, which

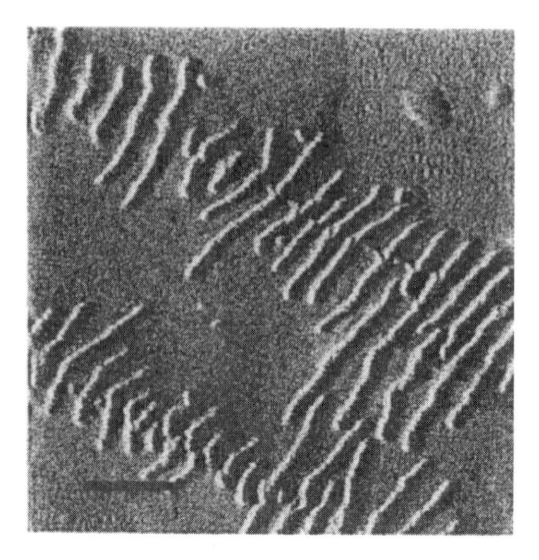

Figure 6.14 An atomic force microscope image of the surface of a polyethylene sample, showing the presence of lamellae; the scale bar represents 300 nm. Reprinted with permission from Hobbs, J. K. and Miles, M. J., *Macromolecules*, **34**, 353–355 (2001). Copyright (2001) American Chemical Society.

shows a map of the surface of a sample of PE, and clearly illustrates the presence of lamellae [18].

6.10 Tribology

Tribology is the term which is used to describe the friction, wear and lubrication of materials. The tribology of polymer systems has become well established in recent years [19]. The *friction* of polymers can be interpreted by using two basic models which describe how work is done at frictional interfaces. A *ploughing* or *deformation component* results from a contact indentor producing a strain, while an *adhesion component* introduces strain due to the interfacial stresses generated by the adhesive forces operating between the two solid surfaces. The adhesive force is given by the following:

$$F = \tau A \tag{6.9}$$

where F is the frictional force, τ is the interfacial shear stress, and A is the contact area. The interfacial shear stress is a function of the contact pressure (P), according to the following:

$$\tau = \tau_0 + \alpha P \tag{6.10}$$

where τ_0 and α are rheological characteristics of the thin film, which can be obtained experimentally. It follows that the friction coefficient (μ) is given by the following:

$$\mu = \tau_0/P + \alpha \tag{6.11}$$

Friction measurements can be made by loading a polymer sample against a rotating shaft, with the frictional force being measured by using transducers.

Wear results as a consequence of friction-induced surface damage. This is a complex phenomenon which begins with the dissipation of frictional energy at the contacting interface. This energy can be dissipated via a number of chemical of mechanical processes. For wear to occur, the debris produced must escape from the contact zone. The rate of wear is described in terms of the mass or volume loss of polymer from the contact zone per unit sliding distance and unit contact pressure.

Lubrication results if polymers that are in contact are separated by a weak interfacial layer (solid or liquid). The extent of the contact will be reduced and both the friction and wear will also be reduced. Lubrication can be classified as *interfacial* (or *boundary*) lubrication when it is based on a thin film, or *fluid* lubrication when based on a thicker film. For some polymer systems in contact, part of one polymer can be transferred to the other surface and then act as a *third-body* lubrication system.

Summary

There are a number of techniques which are capable of providing information about polymer surfaces, ranging from the micron to the angstrom levels. The various techniques that have commonly been used for examining polymer surfaces were introduced in this chapter. These included infrared spectroscopy, Raman spectroscopy, photoelectron spectroscopy, secondary-ion mass spectrometry, inverse gas chromatography, scanning electron microscopy, surface tension measurements, atomic force microscopy and tribological techniques.

References

1. Stuart, B. H., *Modern Infrared Spectroscopy*, ACOL Series, Wiley, Chichester, 1996.
2. Siesler, H. W. and Holland-Moritz, K., *Infrared and Raman Spectroscopy of Polymers*, Marcel Dekker, New York, 1980.
3. Jansen, J. A. J. and Haas, W. E., *Polym. Commun.*, **29**, 77–80 (1988).
4. Ellis, G., Claybourn, M. and Richards, S. E., *Spectrochim. Acta, A*, **46**, 227–242 (1990).
5. Hendra, P., Jones, C. and Warnes, G., *Fourier Transform Raman Spectroscopy: Instrumentation and Chemical Applications*, Ellis Horwood, New York, 1991.
6. Munro, H. S. and Singh, S., 'The Characterization of Polymer Surfaces by XPS and SIMS', in *Polymer Characterization*, Hunt B. J. and James, M. I. (Eds), Blackie, London, 1993, pp. 333–356.
7. Briggs, D., 'Polymer Surface Characterization by XPS and SIMS', in *Characterization of Solid Polymers: New Techniques and Developments*, Spells, S. J. (Ed.), Chapman and Hall, London, 1994, pp. 312–360.

8. Briggs, D., 'Characterization of Surfaces', in *Comprehensive Polymer Science*, Vol. **1**, Booth, C. and Price, C. (Eds), Pergamon Press, Oxford, UK, 1989, pp. 543–560.
9. Briggs, D., Applications of XPS in Polymer Technology, in *Practical Surface Analysis*, Briggs, D. and Seah, M. P. (Eds), Wiley, Chichester, 1990, pp. 437–484.
10. Beamson, G. and Briggs, D., *High Energy XPS of Organic Polymers: The Scienta ESCA 300 Database*, Wiley, Chichester, 1992.
11. Nissen, K. E., Stevens, M. G., Stuart, B. H. and Baker, A. T., *J. Polym. Sci, Polym. Phys. Edn*, **39**, 623–633 (2001).
12. Briggs, D., Brown, A. and Vickerman, J. C., *Handbook of Static SIMS*, Wiley, Chichester, 1989.
13. Lloyd, D. R., Ward, T. C. and Schreiber, H. P., *Inverse Gas Chromatography*, ACS Symposium Series 391, American Chemical Society, Washington, DC, 1989.
14. Thomas, P. S. and Williams, D. R., *Am. Chem. Soc. Symp. Proc., Div. Polym. Mater. Sci. Eng.*, **70**, 416–418 (1994).
15. Tsuji, M., 'Electron Microscopy', in *Comprehensive Polymer Science*, Vol. **1**, Booth, C. and Price, C. (Eds), Pergamon Press, Oxford, UK, 1989, pp. 785–840.
16. Briggs, D., 'Characterization of Surfaces', in *Comprehensive Polymer Science*, Vol. **1**, Booth, C. and Price, C. (Eds), Pergamon Press, Oxford, UK, 1989, pp. 543–559.
17. Cotton, R. J., Engel, A. and Frommer, J. E. (Eds), *Procedures in Scanning Probe Microscopies*, Wiley, Chichester, 1998.
18. Hobbs, J. K. and Miles, M. J., *Macromolecules*, **34**, 353–355 (2001).
19. Briscoe, B. J., *Philos. Mag., A*, **43**, 511–527 (1981).

Chapter 7

Degradation

Learning Objectives

- To appreciate the mechanisms of degradation observed for polymers, including oxidative degradation, thermal degradation, radiation degradation and combustion.
- To understand the dissolution of polymers.
- To use infrared spectroscopy, Raman spectroscopy, electron spin resonance spectroscopy, thermogravimetric analysis, differential scanning calorimetry, thermal mechanical analysis and pyrolysis gas chromatography to study polymer degradation processes.

7.1 Introduction

An understanding of the factors that cause polymer degradation and the mechanisms by which polymers degrade assist in the choice of environment in which these materials are utilized. Degradation affects the appearance and the physical properties of polymers, with some common effects being discoloration and embrittlement. Under extreme conditions, the release of volatile products or even burning may occur. Polymers can degrade due to the rupture of molecular bonds – this type of process may lead to a reduction in molecular weight. Such degradation can result from atmospheric effects and/or from exposure to heat or radiation. Solvents can also be responsible for a deterioration in the polymer mechanical properties. An understanding of how polymers interact with solvents and form solutions is important as such solutions are fundamental to the formulation of paints and adhesives.

A variety of techniques can be used to characterize the degradation mechanisms of polymers and the resulting products that are formed. In this present chapter, the following techniques are discussed: infrared spectroscopy, Raman spectroscopy, electron spin resonance spectroscopy, thermogravimetric analysis, differential scanning calorimetry, thermal mechanical analysis and pyrolysis gas chromatography.

7.2 Oxidative Degradation

Atmospheric components, such as oxygen (O_2) or ozone (O_3), can accelerate chain scission in polymers [1]. Thermoplastics and elastomers are degraded by *peroxidation* in ambient conditions. Peroxidation is a free-radical chain reaction initiated by hydroperoxides:

$$PH + POO^{\bullet} \longrightarrow P^{\bullet} + POOH$$

$$P^{\bullet} + O_2 \longrightarrow POO^{\bullet}$$

where PH represents a polymer molecule. Hydroperoxides are unstable compounds as the peroxide bond undergoes *thermolysis* when heated. This reaction is catalysed by transition-metal ions, which are the main initiators of peroxidation in the absence of light:

$$2POOH \xrightarrow{\text{heat, } M^{+}/M^{2+} \text{ catalyst}} PO^{\bullet} + POO^{\bullet} + H_2O$$

For an outdoor environment, *photolysis* of hydroperoxides is the main initiation process:

$$ROOH \xrightarrow{h\nu,\ M^{+}/M^{2+} \text{ catalyst}} RO^{\bullet} + {}^{\bullet}OH$$

During processing, a polymer may be exposed to high shear forces and this results in *mechanoxidation*:

$$P\text{–}P \xrightarrow{\text{shear}} 2P^{\bullet} \xrightarrow{O_2} 2POO^{\bullet} \xrightarrow{PH} 2POOH$$

where P–P represents a polymer molecule.

Antioxidants are stabilizers added to polymers to inhibit peroxidation. Such additives act by controlling the formation of hydroperoxides. Peroxide decomposers interfere with the fragmentation of hydroperoxides into free radicals, and include compounds containing divalent sulfur or trivalent phosphorus:

$$R_2S + ROOH \longrightarrow R_2S{=}O + ROH$$

$$R_3P + ROOH \longrightarrow R_3P{=}O + ROH$$

Another mechanism by which antioxidants act is by reaction with peroxy radicals. Such compounds reduce the degradation caused by such radicals by 'competing' with the polymer:

$$ROO^{\bullet} + RH \longrightarrow ROOH + R^{\bullet}$$

Substituted phenols and secondary aromatic amines can be used as sources of peroxy radicals. Polymers can also be protected from photooxidation through the process of *radical trapping*. Additives that act via this mechanism are actually light stabilizers rather than antioxidants. Such additives trap both alkyl and peroxy radicals and so interfere with the propagating steps of degradation:

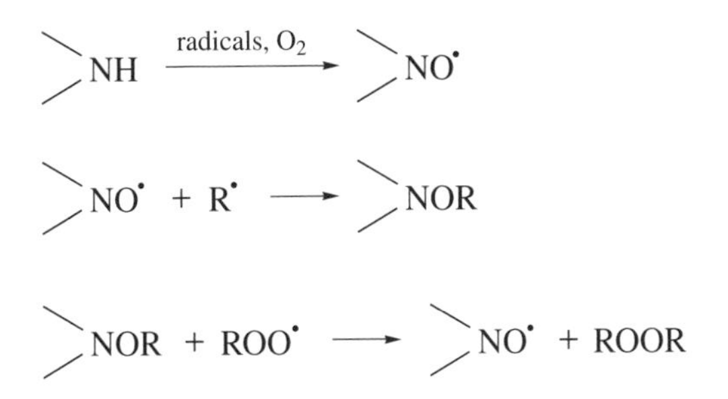

Radical trapping additives are commonly hindered amines.

O_3, formed by the action of ultraviolet light or the electrical discharge on O_2, can also cause significant deterioration of polymers, especially rubber. Low concentrations of ozone are often found in industrial environments and will result in cracking of elastomeric materials. The reaction, known as *ozonolysis*, involves cleavage of the double bond in the rubber backbone to produce two carbonyl compounds:

$$\begin{matrix} R_1 \\ R_2 \end{matrix} \!\!>\! C{=}C \!<\!\! \begin{matrix} R_3 \\ R_4 \end{matrix} \xrightarrow{O_3} \begin{matrix} R_1 \\ R_2 \end{matrix} \!\!>\! C{=}O + O{=}C \!<\!\! \begin{matrix} R_3 \\ R_4 \end{matrix}$$

7.3 Thermal Degradation

The thermal stability of a polymer is related to its bonding energies, e.g. polymers containing higher bonding energies result in more thermally stable materials. For instance, fluorine-containing polymers tend to be the most thermally stable polymers and thus are able to be used at high temperatures. This results from the fact that the magnitude of a C–F bond energy is greater than that of a C–H bond energy, which is in turn greater than that of a C–Cl bond energy.

There are two main types of thermal degradation processes, i.e. *chain depolymerization* and *side-group reactions*. During depolymerization, the polymer chain breaks at some point, leading to reactions in which the products all have the same composition, but consist of smaller molecules. Poly(methyl methacrylate) (PMMA) can degrade thermally via this process:

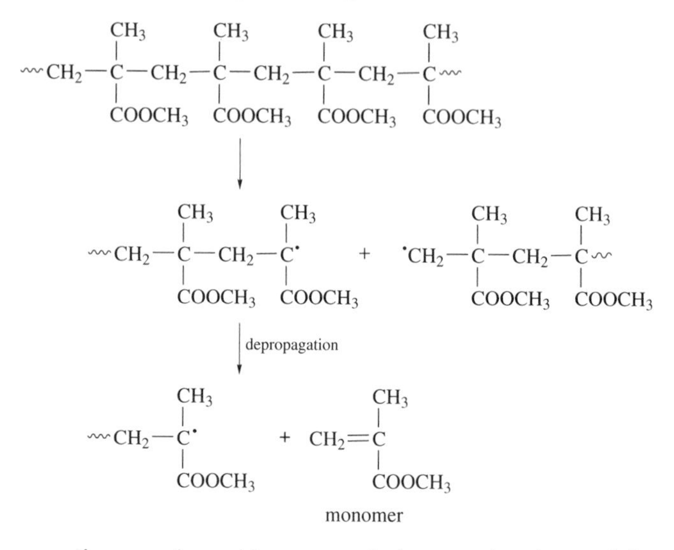

Side-group reactions, such as side-group scission, can be observed for polymers such as poly(vinyl chloride) (PVC):

$$\sim CH_2-\underset{Cl}{CH}-CH_2-\underset{Cl}{CH}-CH_2-\underset{Cl}{CH}-CH_2-\underset{Cl}{CH}-CH_2-\underset{Cl}{CH}\sim$$

$$\downarrow$$

$$\sim CH_2-\underset{Cl}{CH}-CH_2-\underset{Cl}{CH}-CH_2-\underset{Cl}{CH}-CH_2-\underset{Cl}{CH}-CH{=}CH\sim + HCl$$

In this case, one acid molecule has been lost and the double bond formed destabilizes the next repeat unit. This reaction tends to proceed along the chain to produce a conjugated polyene. Cyclization involving adjacent repeat units may also be observed for some polymers. Polymers with COOH side-groups, such as poly(acrylic acid), can degrade via this mechanism (as shown below).

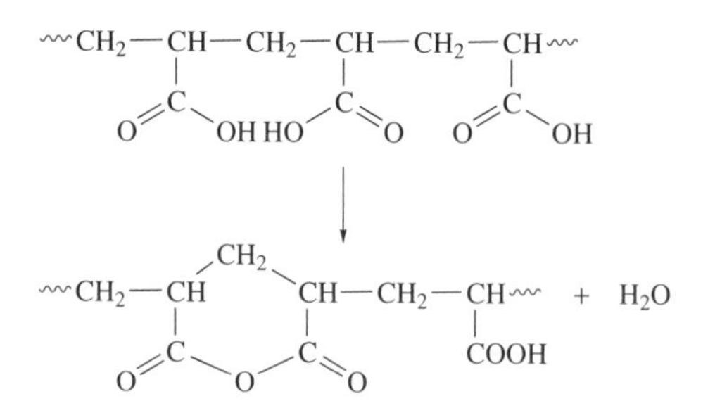

Stabilizers can be used to improve the heat stability of polymers. Traditionally, metals salts and oxides have been used to, for instance, neutralize the hydrogen chloride formed in PVC. Although these materials are effective and inexpensive, they have recently lost favour because they contain heavy metals. Alkyl tin derivatives are now finding acceptance as suitable replacements.

7.4 Radiation Degradation

Some types of radiation, including ultraviolet, X-ray, β- and γ-radiations, are capable of disrupting polymer chains by breaking covalent bonds such as C–Cl, C–C and C–O. Ionization is one process that can occur. As a consequence of the positively charged ion that is produced, a covalent bond is broken and there is a rearrangement of atoms. The breaking of a bond results in scission or cross-linking at the ionization site, depending on the polymer structure and type of radiation. Cross-linking can also be induced by radiation and in some cases may lead to improved physical properties. For instance, PE is cross-linked by using γ-radiation in order to improve its thermal properties.

Stabilizers can be incorporated into polymers to minimize the effect of radiation. The earliest and most effective light stabilizer to be used was carbon black. This additive still finds wide application in tyres, but for aesthetic reasons the white pigment TiO_2 is widely used, for instance, in PVC. TiO_2 reflects white light and absorbs ultraviolet radiation and so inhibits photodegradation processes. Although less effective than inorganic stabilizers, colourless organic light stabilizers, such as 2-hydroxybenzophenone, can be used in conjunction with other antioxidants.

Weathering of polymers is a result of a combination of processes, including the effects of ultraviolet radiation, moisture, temperature and thermal cycling. As many polymers find use in outdoor applications, it is important to have an understanding of the weathering properties of such materials.

7.5 Combustion

Many *pure* polymers are flammable and this property needs to be minimized, particularly in applications such as textiles and toys. The combustion of polymers is initiated by a source of heat that causes the temperature of the polymer to increase to a level where the chemical bonds begin to break. Low-molecular-weight products are formed at this stage and such products can migrate out of the polymer and disappear into the gas phase. A gas-phase oxidation then occurs and the heat that is produced further heats the polymer, thus causing an additional breakdown of the polymer. This process continues until no further volatile products are produced by the polymer. The combustion behaviour of polymers may be quantified by determining the *limiting oxygen index* (LOI) of the material. The technique involves combusting a polymer sample of a specific size with varying oxygen–nitrogen mixtures to determine the minimum amount required to allow combustion for a minimum period of three minutes.

The flammability resistance of polymers can be improved by the addition of *flame retardants*. The latter act in a number of ways, depending on the chemical nature of the additive. They may cool the system, provide a barrier by charring, or quench chain reactions. The most common flame retardant used for thermoplastics is antimony trioxide, although phosphate esters may also be used to reduce flammability.

7.6 Dissolution

When certain polymers come into contact with particular solvents, *swelling* of the polymer can occur [2]. Swelling is caused by the breaking of weak intermolecular bonds by the solvent molecules. The polymer chains move apart and the material becomes flexible (e.g. PVAl in water). Swelling can be minimized by employing cross-linked or by crystalline polymers, or by the use of incompatible polymer and solvent systems.

Dissolving a polymer in a solvent is a slow process which occurs in two stages: first, the solvent molecules slowly diffuse into the polymer to produce a swollen gel, and then polymer–solvent interactions cause the gel to gradually disintegrate. Solubility occurs when the free energy of mixing (ΔG) is negative, as follows:

$$\Delta G = \Delta H - T\Delta S < 0 \tag{7.1}$$

The heat of mixing per unit volume (ΔH) is approximated as follows:

$$\Delta H = v_1 v_2 (\delta_1 - \delta_2)^2 \tag{7.2}$$

where v_1 is the volume fraction of the solvent, v_2 is the volume fraction of the polymer, δ_1 is the solubility parameter for the solvent and δ_2 is the solubility parameter for the polymer. In the absence of a strong interaction, such as hydrogen bonding, solubility is expected if $(\delta_1 - \delta_2)$ is less than values in the range 3.4–4.0 $J^{1/2}\ cm^{-3/2}$. Solubility parameters can be calculated for both polymers and solvents, with Table 7.1 listing the values measured for some common systems.

Table 7.1 Solubility parameters for some common polymers and solvents

Polymer/solvent	Solubility parameter ($J^{1/2}$ $cm^{-3/2}$)
PE	16.2
PP	16.6
PS	17.6
PMMA	18.6
PVC	19.4
Nylon 6,6	27.8
n-Hexane	14.8
Carbon tetrachloride	17.6
Benzene	18.7
Acetone	19.9
Water	47.9
Methanol	29.7

SAQ 7.1

Determine, using the solubility parameters given in Table 7.1, whether polyethylene is soluble in *n*-hexane.

In a polymer solution, two segments of a polymer molecule cannot occupy the same space, and therefore will experience increasing repulsion as they move closer together. Thus, a polymer has a surrounding region where its segments cannot move and this region is called the *excluded volume*. The latter contributes to an excess in entropy of the solution. In addition, the attractions and repulsions that occur between polymer molecules contribute to an excess in enthalpy. For most polymer–solvent systems, there is a unique temperature at which these effects cancel out. This temperature is known as the *Flory temperature* or *theta temperature* (θ).

Solvents for a particular polymer may be classified on the basis of the θ temperatures for that polymer. Solvents are described as *good* if θ lies well below room temperature. In a good solvent, the polymer becomes well solvated by the solvent molecules and the conformation of its molecules expands. A solvent is described as *poor* if θ is above room temperature. In a poor solvent, the polymer is not well solvated and adopts a constricted conformation.

When the temperature of a polymer solution is raised or lowered, the solvent becomes *thermodynamically poorer*. A temperature is reached beyond which the polymer and solvent are no longer miscible, and at extreme temperatures the mixture separates into two phases. A phase diagram of a polymer solution is shown in Figure 7.1. This diagram illustrates phase separation both on heating and on cooling. When the temperature is increased, the *lower critical solution temperature*

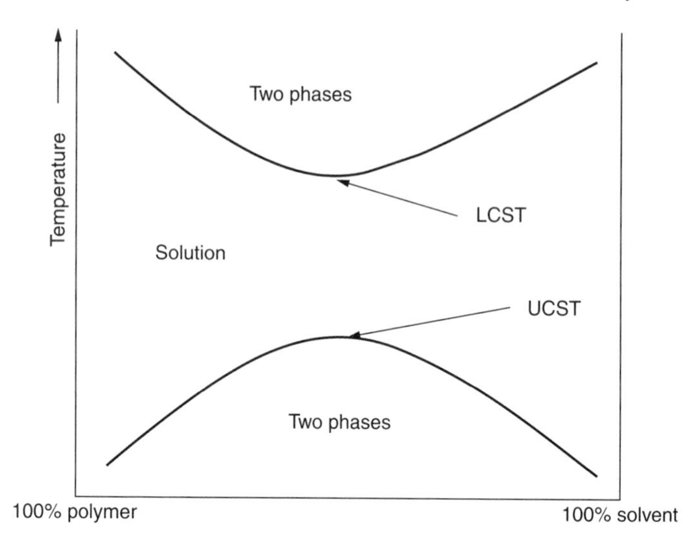

Figure 7.1 A phase diagram of a polymer solution, showing the positions of the lower critical solution and upper critical solution temperatures.

(LCST) is reached, with this temperature marking a transition from one phase to two phases. When the temperature of the polymer solution is reduced, the *upper critical solution temperature* (UCST) is reached, with this representing the lower-temperature phase transition.

7.7 Infrared Spectroscopy

Infrared spectroscopy can be used to investigate the effects of environmental exposure of polymers. The exposure to ultraviolet radiation, chemical attack by free radicals produced from natural processes, and microbial attack can all be studied by using infrared spectroscopy. The role of this technique in polymer degradation is illustrated by its application to thermally and photooxidized polyethylenes [3]. During the heat-oxidation process with PE, a range of carbonyl-containing compounds are formed. These decomposition products give rise to a broad C=O stretching band at about 1725 cm^{-1}, consisting of a number of overlapping component bands. When the oxidized samples are treated with an alkali, a shoulder at 1715 cm^{-1} disappears and is replaced by a distinctive peak near 1610 cm^{-1}. This band is due to C=O stretching of the COO^- ion of a salt, indicating that the shoulder at 1715 cm^{-1} is characteristic for saturated carboxylic acids. Another shoulder at 1735 cm^{-1} is characteristic of a saturated aldehyde. However, the major contribution to the carbonyl band is due to the presence of saturated ketones. The broad C=O stretching band is also present in the infrared spectrum of photooxidized PE samples, which also show additional bands at

990 and 910 cm^{-1}. The latter bands are characteristic of vinyl groups and their presence shows that chain-terminating unsaturated groups are being formed, most likely as a result of chain scission.

7.8 Raman Spectroscopy

Fourier-transform (FT) Raman spectroscopy has been used as a means of investigating the solvent-induced softening of poly(ether ether ketone) (PEEK) [4]. 1,1,2,2-Tetrachloroethane (TCE) is readily absorbed by amorphous PEEK, showing a mass uptake of 165% after exposure for 24 h, and causes the clear amorphous polymer to become opaque in appearance like the crystalline material. The FT Raman spectrum of amorphous PEEK after exposure to TCE shows a shift in frequency to the C=O stretching mode and a change in the relative intensity of the symmetrical C–O–C stretching mode. While the C=O stretching mode appears at 1651 cm^{-1} in the spectrum of the untreated amorphous material, after exposure to solvent it shifts to 1644 cm^{-1}, a frequency also observed for the crystalline material. The effect may be explained by Lewis acid–base interactions between the polymer and the solvent. PEEK acts as a soft base due to the presence of the C=O, C–O–C and aromatic groups in the structure, which can act as electron donors, while the TCE may act as an acid. The swelling of PEEK occurs because the interaction of the solvent with the polymer results in dissolution of some of the solvent in the polymer phase as well as the dissolving of some of the polymer in the solvent phase. At the same time, crystallinity is induced in the polymer by the solvent. The interaction between PEEK and the chlorinated solvent causes electronic charge to be withdrawn from the C=O bond of PEEK into the polymer–solvent bond. This lowers the C=O bond force constant and causes a shift to lower frequencies of vibration of the C=O stretching mode compared to the amorphous environment.

7.9 Electron Spin Resonance Spectroscopy

Electron spin resonance (ESR) spectroscopy has been successfully employed in the study of degradation mechanisms in polymers [5]. This technique is able to detect the presence and nature of radical degradation products. For example, when polydimethylsiloxane (PDMS) is exposed to γ-radiation at room temperature, the ESR spectrum shows a singlet that can be attributed to the following radical [6]:

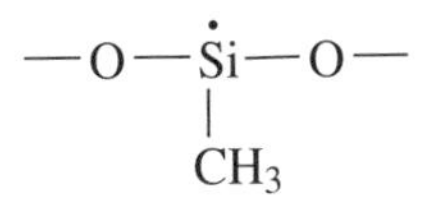

When PE is irradiated at 77 K, an alkyl radical is formed [5]:

$$\underset{\gamma}{-CH_2}-\underset{\beta}{CH_2}-\underset{\alpha}{\dot{C}H}-\underset{\beta}{CH_2}-\underset{\gamma}{CH_2}-$$

This radical is observed as a six-line ESR spectrum. In the radical, the unpaired electron occupies a carbon p-orbital, and only the protons in the α- and β-positions have measurable coupling constants.

DQ 7.1

When PVC is exposed to ultraviolet light, the following free radical is initially formed:

$$\left[CH_2\underset{\underset{Cl}{|}}{C}H \right] \xrightarrow{UV} \left[CH_2\underset{\bullet}{C}H \right] + Cl^{\bullet}$$

The ESR spectrum of PVC exposed to ultraviolet radiation is shown below in Figure 7.2 [7]. Attribute the hyperfine splitting displayed in this figure to the structure of the radical which is formed.

Answer

Six lines are observed in the ESR spectrum as a result of the interaction of the unpaired electron with the five surrounding protons.

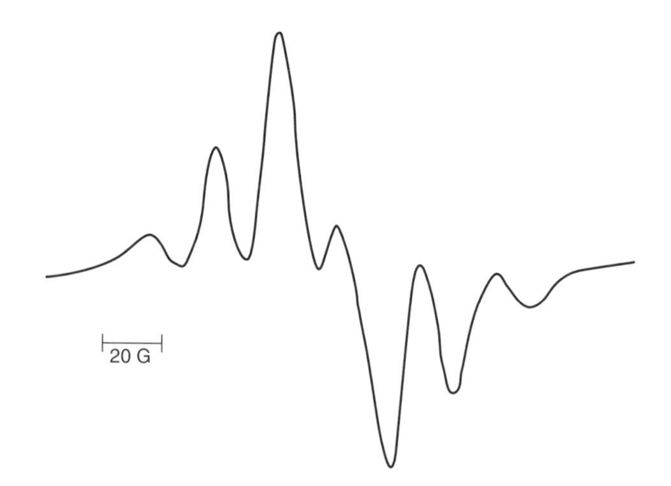

Figure 7.2 The ESR spectrum of a sample of poly(vinyl chloride) after exposure to ultraviolet radiation (DQ 7.1). From Yang, N. L., Liutkas, J. and Haubenstock, H., *Polymer Characterization by ESR and NMR*, Woodward, A. E. and Bovey, F. A. (Eds), ACS Symposium Series 142, pp. 35–48 (1980). Copyright (1980) American Chemical Society.

7.10 Thermal Analysis

7.10.1 Thermogravimetric Analysis

Thermogravimetric analysis (TGA) is a thermal method that involves the measurement of weight loss as a function of temperature or time [8, 9]. TGA can be used to quantify the mass change in a polymer associated with transitions or degradation processes. TGA data provide characteristic curves for a given polymer because each polymer will show a unique pattern of reactions at specific temperatures.

In TGA, a sample is placed in a furnace while being suspended from one arm of a sensitive balance. The change in sample mass is recorded while the sample is maintained either at a required temperature or while being subjected to a programmed heating sequence. The layout of a typical TGA system is shown schematically in Figure 7.3. The thermobalance can detect to 0.1 μg, and calibration can be made by using standard masses. Calcium oxalate monohydrate ($CaC_2O_4.H_2O$) is commonly used for calibration. A heater allows temperatures up to 2800°C to be obtained, and temperature rates can be varied from 0.1 to 300°C min^{-1}, with both heating and cooling of the samples being possible. Reactions studied in the thermobalance can be carried out under different atmospheres, such as nitrogen, argon, helium or oxygen.

The TGA curve can be plotted as the sample mass loss as a function of temperature, or alternatively, in a differential form where the change in sample mass with time is plotted as a function of temperature. The different forms of these curves are illustrated in Figure 7.4, which shows the results of a TGA experiment

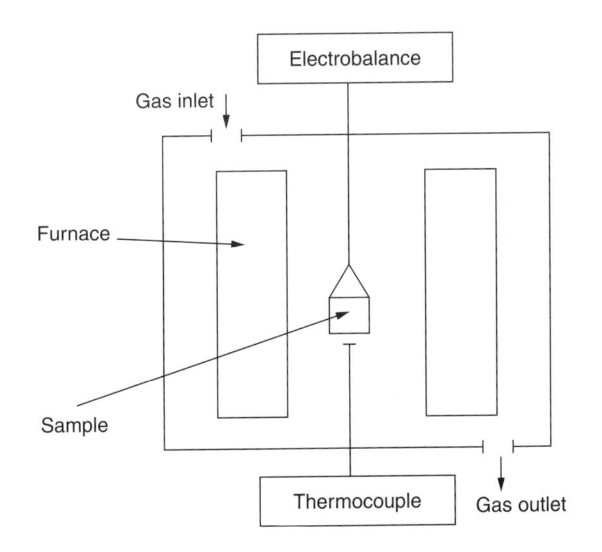

Figure 7.3 Schematic of a typical thermogravimetric analysis instrument.

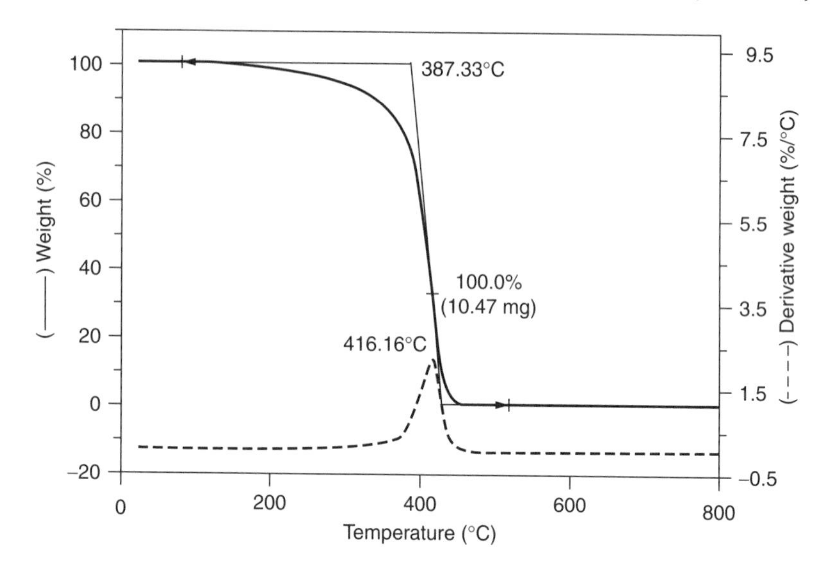

Figure 7.4 Thermogravimetric analysis data obtained for a sample of polystyrene over the range from room temperature to 800°C, illustrating the different types of curves which can be obtained. From Sandler, S. R., Karo, W., Bonesteel, J. and Pearce, E. M., *Polymer Synthesis and Characterization: A Laboratory Manual*, © Academic Press, 1998. Reproduced by permission of Academic Press.

on a polystyrene (PS) sample from room temperature to 800°C. The TGA curves of polymers can also be used for qualitative determination. Figure 7.5 shows the TGA curves obtained for a number of polymers and illustrates that different polymers can be distinguished by comparison of variables such as the temperature range or the activation energy of decomposition.

SAQ 7.2

Using the TGA data supplied in Figure 7.5, determine which of the polymers shown is the least thermally stable.

Polymers can exhibit a wide range of degradation processes. For instance, polymers such as PMMA and PS depolymerize, while PE will produce unsaturated hydrocarbons from chain segments of varying lengths. The temperatures of degradation of polyalkenes are affected by substitution. For instance, PTFE decomposes at a much higher temperature than PE because of the fluorine substitution. In comparison, polypropylene decomposes at a lower temperature than polyethylene because of the substitution of a methyl group. PVC and polyacrylonitrile (PAN) can eliminate small molecules initially and form unsaturated links and cross-linking before eventually degrading via complex reactions to

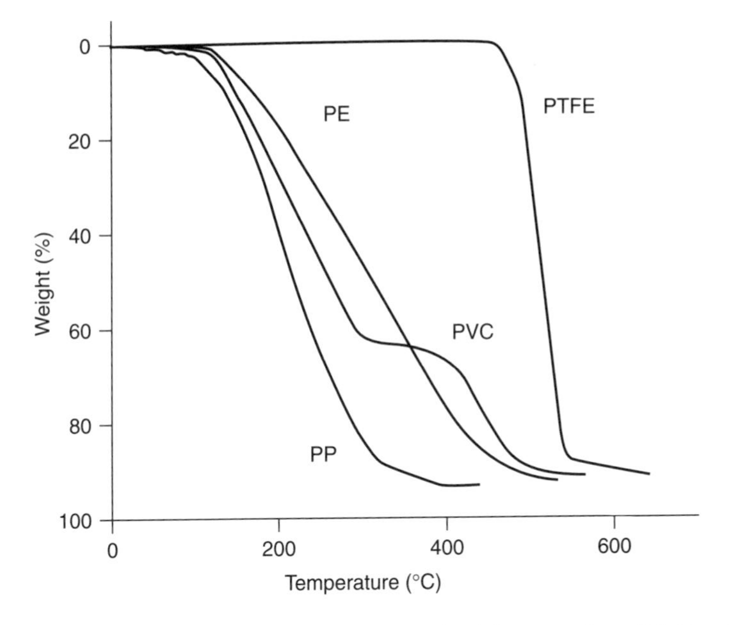

Figure 7.5 Thermogravimetric analysis curves obtained for a number of different polymers, illustrating how such data can be used for qualitative determination.

chars which will oxidize in air. Polyamides can absorb moisture, the subsequent loss of which can be observed below 100°C, sometimes in stages. Polymers such as cellulose, polyester resins and phenol–formaldehyde resins display complex decomposition schemes. Such decomposition processes often eliminate small molecules that can be flammable or toxic. These reactions are further complicated in an oxidizing atmosphere.

Figure 7.6 illustrates how TGA can be used to determine the mass loss. The loss associated with an initial step, such as solvent evaporation, is given by $(w_0 - w_1)$. This curve shows two degradation processes – the second step represents the first degradation process, where the mass loss is given by $(w_1 - w_2)$, while the third step, representing a further degradation process, shows a mass loss of $(w_2 - w_f)$, where w_f is the residue that does not decompose in the temperature range covered by the experiment. The derivative curve (DTG) shows a peak associated with each separate step, which represents the maximum rate of mass loss.

SAQ 7.3

Figure 7.7 below shows the TGA results obtained for a sample of PVC heated from room temperature to 800°C in a N_2 atmosphere. Discuss the changes that the polymer undergoes as it decomposes over this temperature range.

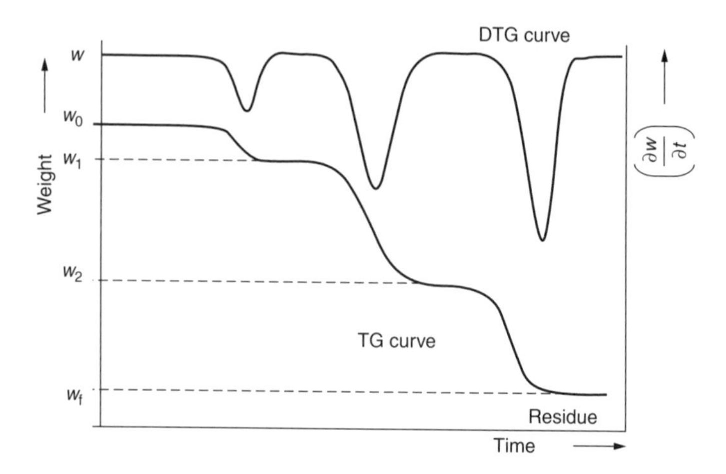

Figure 7.6 Illustration of how thermogravimetric analysis can be used to investigate mass losses in a (polymer) system.

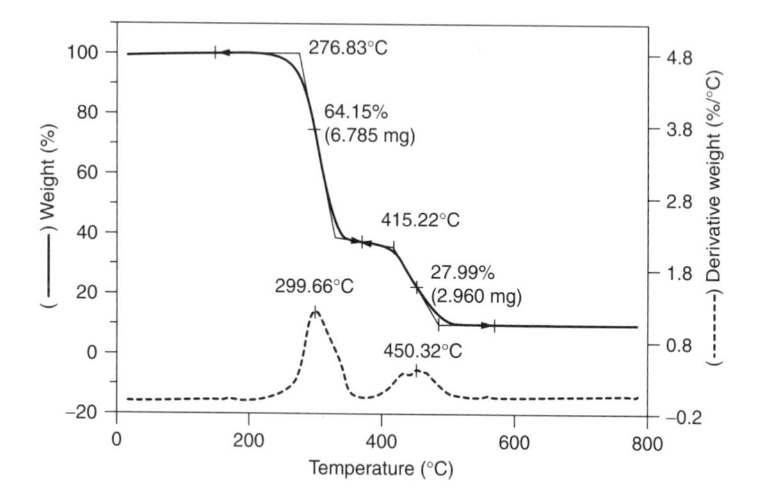

Figure 7.7 Thermogravimetric analysis data obtained for a sample of poly(vinyl chloride) over the range from room temperature to 800°C in a N_2 atmosphere (SAQ 7.3). From Sandler, S. R., Karo, W., Bonesteel, J. and Pearce, E. M., *Polymer Synthesis and Characterization: A Laboratory Manual*, © Academic Press, 1998. Reproduced by permission of Academic Press.

TGA can be used to examine polymer mixtures by showing differences between the behaviours of the individual substances on heating. If significant temperature differences are observed, then the specific reactions of the particular components can be identified. For instance, polymer blends, copolymers, and polymers containing additives can all be examined in this way. For phthalate

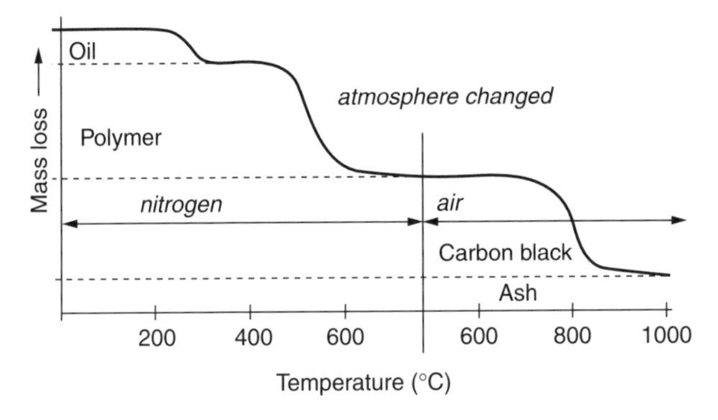

Figure 7.8 Thermogravimetric analysis data obtained for an elastomer containing carbon black.

plasticizers, a sharp peak near 300°C in the derivative TGA curve due to evaporation of the plasticizer will be present. In the case of a $CaCO_3$ filler, the TGA curve shows a decomposition to CaO near 800°C. Other fillers, such as $Al_2O_3.3H_2O$, eliminate the water of hydration to give Al_2O_3, or they can be oxidized to higher oxidation states, such as PbO to PbO_2. Figure 7.8 shows the TGA curves obtained for an elastomer containing carbon black. The analysis of the latter is carried out by an initial degradation of the other organic components in the system in an inert N_2 atmosphere, followed by oxidation of the carbon black in O_2. Inorganic fillers will remain at the end of the scan. Note that the oxidation of carbon must occur over a temperature range where no other degradation occurs.

7.10.2 Differential Scanning Calorimetry

Differential scanning calorimetry (DSC) can be used to study the oxidative degradation of polymers [9]. A standard test involves heating a sample in N_2 to 200°C, at which point the atmosphere is changed to O_2 with the temperature held constant (at 200°C). The time for the onset of the exothermic oxidation is recorded. As an alternative, the polymer sample can be heated in O_2 and the temperature at which the onset of oxidation occurs is reported. Figure 7.9 illustrates the standard and isothermal DSC curves obtained for the oxidation of a sample of polyethylene.

7.10.3 Thermal Mechanical Analysis

Thermal mechanical analysis (TMA) is a useful method for the investigation of solvent swelling of polymers. The changes in sample dimensions during absorption and desorption can be measured by using this technique. An evaluation of

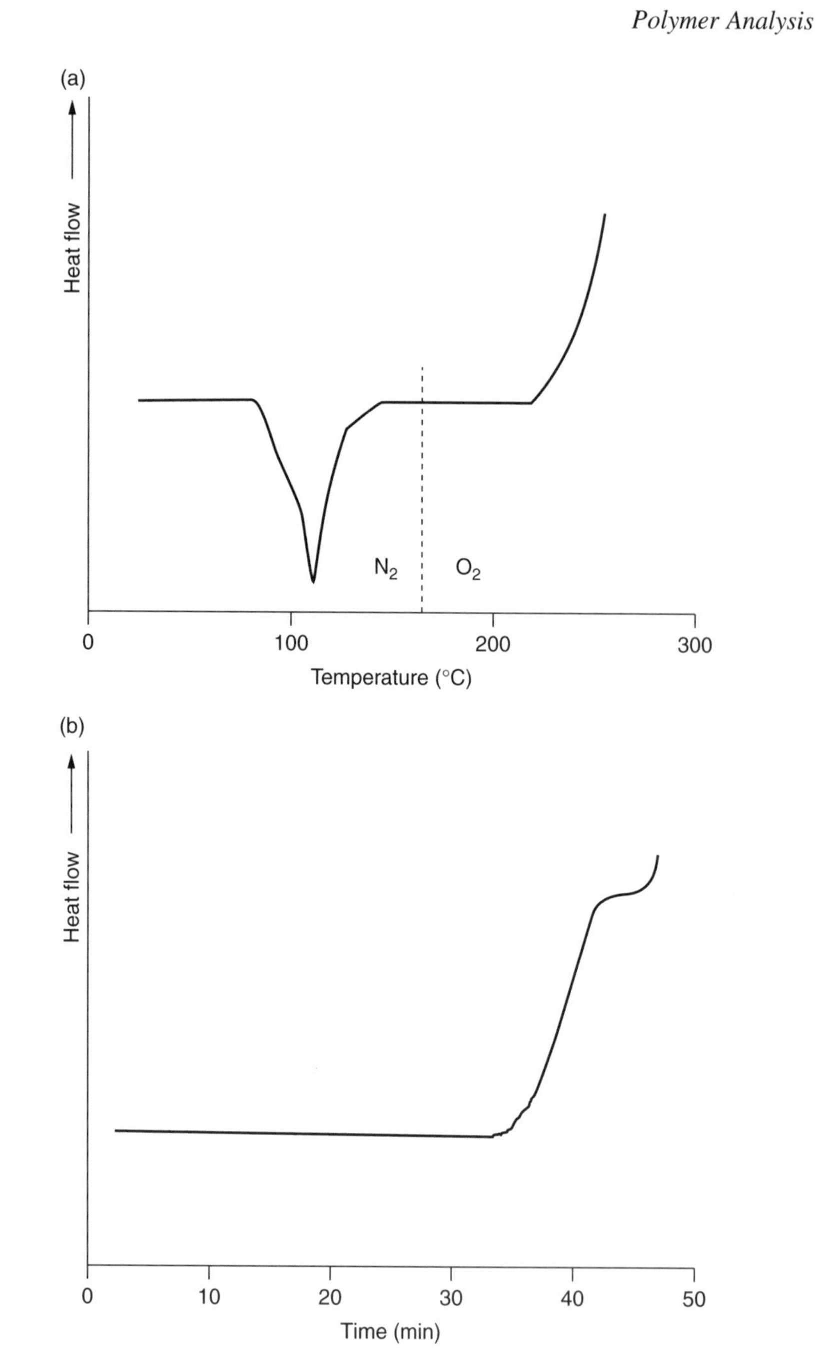

Figure 7.9 Differential scanning calorimetry curves obtained for the oxidation of a sample of polyethylene: (a) standard conditions; (b) isothermal conditions.

the swelling behaviour may be used, for example, to determine the cross-link density of certain polymers.

7.11 Pyrolysis Gas Chromatography

Pyrolysis gas chromatography (PGC) is a technique which involves the thermal degradation of a sample at temperatures above 400°C [10]. In this method, samples are heated in an inert atmosphere either by furnace chamber pyrolysis,

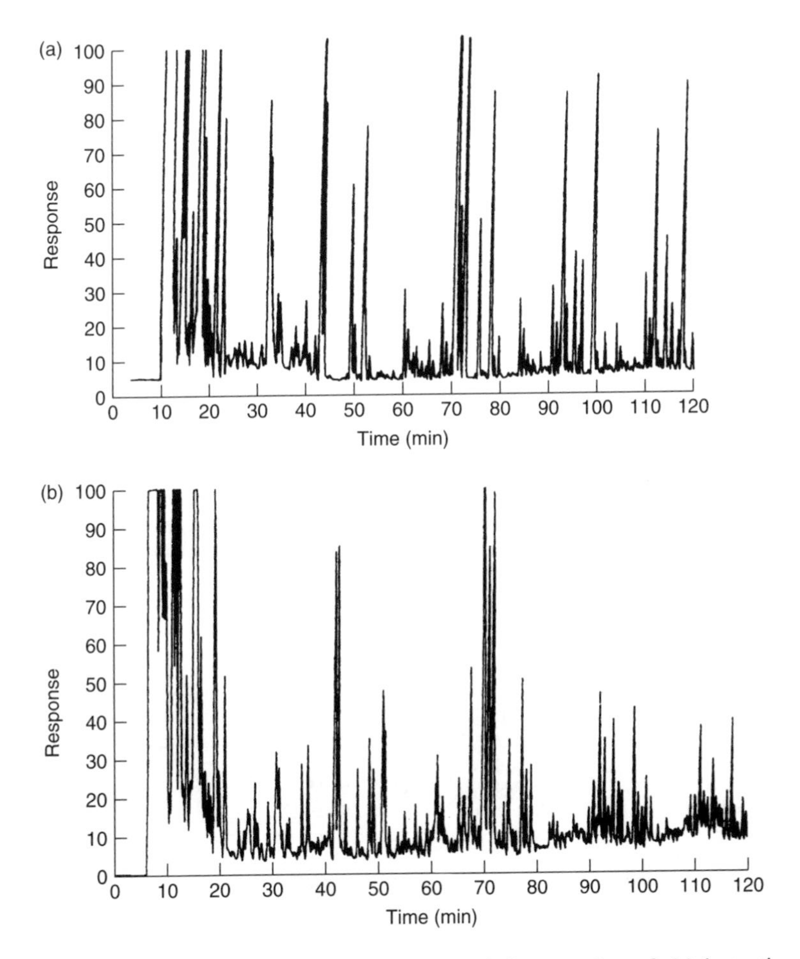

Figure 7.10 Pyrolysis gas chromatograms obtained for samples of (a) isotactic and (b) atactic polypropylene. Reproduced from S.A. Liebman and E.V. Levy (eds), *Pyrolysis and GC in Polymer Analysis*, Marcel Dekker, Inc. N.Y., 1985, P4, by courtesy of Marcel Dekker, Inc.

flash pyrolysis or laser pyrolysis techniques. The fragments formed are more volatile than the initial polymer and can be separated by an attached gas chromatograph. Mass spectrometry can also be used to identify the peaks eluted from the chromatograph.

The pyrogram produced by this method acts as a characteristic fingerprint and databases containing such information have been collected. The data provided by PGC is useful for both the qualitative and quantitative analysis of polymers. Figure 7.10 provides an example of the complex and distinct fragmentation patterns produced by this technique, in this case showing the pyrograms of isotactic and atactic polypropylene samples.

Summary

The types of degradation mechanisms observed in polymers were introduced in this chapter. Polymer solutions were also discussed. The techniques that can be used to understand and evaluate polymer degradation processes, including infrared spectroscopy, Raman spectroscopy, electron spin resonance spectroscopy, thermogravimetric analysis, differential scanning calorimetry, thermal mechanical analysis and pyrolysis gas chromatography, were also outlined.

References

1. Carlsson, D. J. and Wiles, D. M., 'Degradation', in *Encyclopedia of Polymer Science and Engineering*, Vol. 4, Mark, H. F. (Ed.), Wiley, New York, 1987, pp. 630–696.
2. Casassa, E. F. and Berry, G. C., 'Polymer Solutions', in *Comprehensive Polymer Science*, Vol. 2, Booth, C. and Price, C. (Eds), Pergamon Press, Oxford, UK, 1989, pp. 71–120.
3. Bower, D. I. and Maddams, W. F., *The Vibrational Spectroscopy of Polymers*, Cambridge University Press, Cambridge, UK, 1989.
4. Stuart, B. H. and Williams, D. R., *Polymer*, **35**, 1326–1328 (1994).
5. Ranby, B. and Rabek, J. F., *ESR Spectroscopy in Polymer Research*, Springer-Verlag, Berlin, 1977.
6. Ormerod, M. G. and Charlesby, A., *Polymer*, **4**, 459–464 (1963).
7. Yang, N. L., Liutkas, J. and Haubenstock, H., 'An ESR Study of Initially Formed Intermediates in the Photodegradation of Poly(Vinyl Chloride)', in *Polymer Characterization by ESR and NMR*, Woodward, A. E. and Bovey, F. A. (Eds), ACS Symposium Series 142, American Chemical Society, Washington, DC, 1980, pp. 35–48.
8. Sandler, S. R., Karo, W., Bonesteel, J. and Pearce, E. M., *Polymer Synthesis and Characterization: A Laboratory Manual*, Academic Press, San Diego, CA, 1998.
9. Haines, P. J., *Thermal Methods of Analysis: Principles, Applications and Problems*, Blackie, London, 1995.
10. Liebman, S. A. and Levy, E. V. (Eds), *Pyrolysis and GC in Polymer Analysis*, Marcel Dekker, New York, 1985.

Chapter 8

Mechanical Properties

Learning Objectives

- To understand the stress–strain, viscous, viscoelastic and elastic behaviours exhibited by polymer systems.
- To appreciate the different methods of processing polymers.
- To characterize the mechanical properties of polymers by using tensile, flexural, tear, fatigue, impact and hardness tests.
- To use viscometry to measure the viscosity of polymers.
- To use dynamic mechanical analysis to understand the viscoelastic behaviour of polymers.

8.1 Introduction

The mechanical behaviour of amorphous polymers can be classified according to different molecular mechanisms [1]. Viscous flow may be observed, which involves the irreversible bulk deformation of a polymer. This mechanism is associated with the irreversible slippage of molecular chains past one another. Rubber-like elasticity is observed in polymers where there is small-scale movement of chain segments, although large-scale movement such as flow is prevented by a network structure. There also exists the case where a polymer may flow in response to an applied stress, and the extent of this flow is time-dependent. Such behaviour may be regarded as a combination of elastic and viscous responses and is known as *viscoelasticity*. Hookean elasticity may also be observed where the polymer behaves as a glass. In this case, the motion of polymer chains is restricted, with only bond stretching and bending being observed. The mechanical properties of semicrystalline polymers cannot be described so simply. Their properties will depend on the nature of the crystalline regions. However, the rheological behaviour of crystalline polymers may be inferred in part from the behaviour of amorphous polymers.

There are a number of fundamental techniques used to characterize the mechanical properties of polymers, including tensile, flexural, tear strength, fatigue, impact and hardness tests. All of these are examined in this present chapter, along with viscometry and dynamic mechanical analysis which can be used to characterize the flow properties of polymers. The common methods for processing polymers are also discussed.

8.2 Stress–Strain Behaviour

The degree to which a material will strain depends on the magnitude of the imposed stress [2]. This *stress* (σ) is defined as the load (F) per unit area (A), given as follows:

$$\sigma = F/A \tag{8.1}$$

There are three basic types of stress measurements – *tensile stress* is the resistance of a material to stretching forces, *compressive stress* is the resistance of a material to 'squashing' forces, while *shear stress* is the resistance of a material to 'push–pull' forces.

The strain (ε) is the amount of deformation per unit length of the material due to the applied load, and is given as follows:

$$\varepsilon = (l_i - l_0)/l_0 = \Delta l/l_0 \tag{8.2}$$

where l_0 is the original length of the sample before any load is applied, l_i is the instantaneous length, and Δl is the amount of elongation.

Deformation where the stress and strain are proportional is called *elastic deformation*. In such a case, a plot of stress against strain produces a linear graph. The slope of such a plot provides the *Young's modulus* (also known as the *modulus of elasticity* or the *tensile modulus*) (E) of the material, a proportionality constant. Typical values of Young's moduli for some common polymers are presented in Table 8.1. The modulus can be thought of as the 'stiffness' of a polymer. The parameter E can be evaluated from the slope of the linear elastic portion of a force–extension curve where:

$$E = (\text{slope} \times \text{gauge length})/(\text{cross-sectional area}) \tag{8.3}$$

For many polymers, the initial elastic region is not linear and so it is not possible to determine Young's modulus from the slope. However, a tangent or secant modulus may be used. The latter modulus is often taken as the slope of the stress–strain curve at a specified strain (usually 0.2%). This type of analysis will be discussed below in Section 8.7.

There are three general types of stress–strain behaviours observed for polymers and these are illustrated in Figure 8.1. Brittle polymers deform elastically

Table 8.1 Tensile properties of some common polymers and polymer composites

Polymer	Tensile strength (MPa)	Young's modulus (GPa)
LDPE	5–25	0.1–0.3
HDPE	15–40	0.5–1.2
UHMWPE	20–40	0.2–1.2
PP	25–40	0.9–1.5
PVC (unplasticized)	25–70	2.5–4.0
PS	30–100	2.3–4.1
PMMA	80	2.4–3.3
Nylon 6,6	82	3.3
Nylon 6	78	2.6–3.0
Nylon 11	1.5	44
Nylon 6,6 (with 30% glass fibre)	160–210	10–11
PTFE	10–40	0.3–0.8
PTFE (with 25% glass fibre)	7–20	1.7
PVDC	25–110	0.3–0.6
PVDF	25–60	1.0–3.0
Cellulose acetate	12–110	1.0–4.0
PET	80	2.4
PBT	50	2.0
Phenolic resins	35–62	2.8–4.8
Epoxy resin	28–90	2.4
Epoxy resin (with 50% glass fibre)	800	30
Epoxy resin (with 50% carbon fibre)	1000	80
Polyester resins	41–90	2.1–4.4
Silicone rubber	6.5	62
PC	55–75	2.3–2.4
PPO	55–65	2.5
Polyimides	70–150	2.0–3.0
Kevlar	2760	59–124
PEEK	70–100	3.7–4.0
SBR	12–21	0.0021–0.010
Nitrile rubber	7–24	0.007–0.024
ABS	41–45	2.1–2.4

(curve A), while elastomeric polymers show a deformation that is totally elastic (curve B). These polymers exhibit large recoverable strains produced at low stress levels. Elastic deformation occurs when the stress and strain are proportional and a plot of stress versus strain is linear. Plastic polymers initially display elastic deformation in stress–strain studies, although this type of deformation is followed by yielding and a region of plastic deformation (curve C). The latter represents a permanent change, and the stress is no longer proportional to the strain.

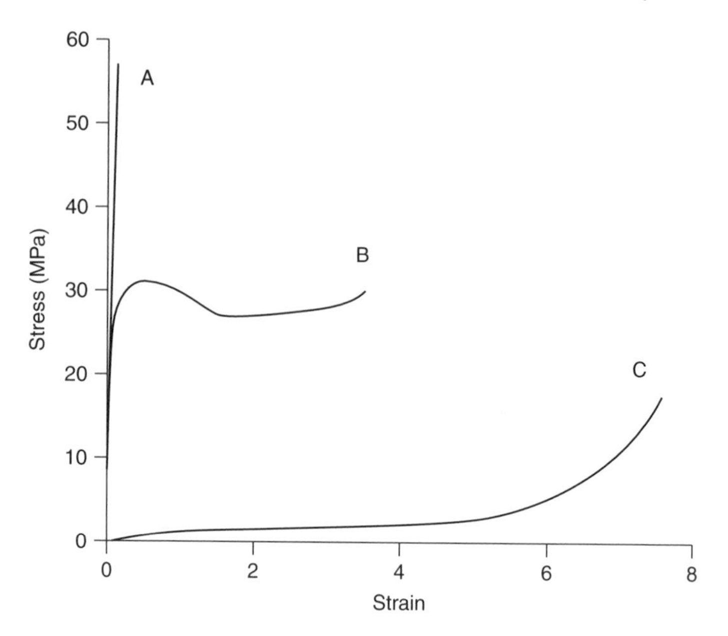

Figure 8.1 The three general types of stress–strain behaviours exhibited by polymeric materials: (A) brittle polymers; (B) elastomeric polymers; (C) plastic polymers.

The onset of plastic deformation and the stress at the maximum of this plot (curve C) is known as the *yield strength* (σ_Y), while the stress at which the fracture of the material occurs is known as the *ultimate tensile strength* (UTS). The latter corresponds to the maximum stress that can be sustained by the material in tension – if this stress is maintained, then fracture will result. The tensile strengths of some common polymers are given in Table 8.1.

DQ 8.1

What differences would be observed when comparing the tensile plots of a brittle polymer and a ductile polymer?

Answer

A brittle polymer will show an initial linear elastic region on a force–extension plot. In comparison, a ductile polymer, such as a soft thermoplastic, will just display a continuous curve.

The degree of elongation of a polymer provides a measure of the ductility of the material. As there is usually a large amount of elastic recovery when a thermoplastic sample breaks, the *percentage elongation* at break is often quoted.

The percentage elongation is given by the following:

$$\% \text{ elongation} = (\text{increase in gauge length} \times 100)/(\text{original gauge length}) \quad (8.4)$$

Thermoplastics subjected to continuous stress above the yield point experience the process of *cold-drawing*. At the yield point, the sample forms a 'neck'. Once the polymer is fully cold-drawn, it is stronger than during neck propagation, due to the alignment of the polymer molecules. This explains the final upswing observed in the stress–strain curve.

For composites containing long parallel fibres, the Young's modulus (E_c) in the fibre direction is given by the following:

$$E_c = E_f V_f + E_m V_m \quad (8.5)$$

where E_f and E_m are the Young's moduli of the fibre and matrix, respectively, and V_f and V_m are, respectively, the volume fractions of the fibre and matrix. The tensile strength (σ_c) in the fibre direction is given as follows:

$$\sigma_c = \sigma_f V_f + \sigma_m V_m \quad (8.6)$$

where σ_f and σ_m are the tensile strengths of the fibre and matrix, respectively. The tensile strengths and Young's moduli for some common polymer composites and their components are listed in Table 8.1.

SAQ 8.1

A fibre-reinforced composite material is made from a polyester resin and long uniaxially aligned glass fibres, with a fibre volume fraction of 45%. Estimate the Young's modulus and the tensile strength of this composite.

8.3 Viscous Flow

Viscous flow is the irreversible bulk deformation of a polymer associated with the irreversible slippage of the molecular chains past one another [1]. *Shear stress* (τ) is the stress originating from viscous motion and is defined as the ratio of the force (F) tangentially applied to the flowing surface area (S), as follows:

$$\tau = F/S \quad (8.7)$$

The *shear rate* ($\dot{\gamma}$) is defined as the gradient of the velocity (v) of the flowing liquid, as follows:

$$\dot{\gamma} = \mathrm{d}v/\mathrm{d}x \quad (8.8)$$

where x is the direction perpendicular to the flow. The *viscosity* (η) is the ratio between the shear stress and the shear rate, according to the following:

$$\eta = \tau/\dot{\gamma} \quad (8.9)$$

If η is independent of the shear rate, then the liquid is described as *Newtonian*. Thus, when the shear rate is plotted against the shear stress, the graph is linear. In many cases, $\eta(= \tau/\dot{\gamma})$ is not constant, and the relationship is then described by a power law, as follows:

$$\tau = K\dot{\gamma}^n \tag{8.10}$$

where n is the *melt flow index* and is a measure of *non-Newtonian* behaviour, and K is a proportionality constant.

The melt viscosity of a polymer is dependent on the molecular weight of the sample. A log–log plot of viscosity as a function of the chain length shows two distinct regions, depending upon whether the chains are long enough to be significantly entangled (Figure 8.2). In this figure, Z_w is the *weight-average chain length*, and for chain lengths below some critical value (Z_c) – usually about 600 for most polymers – the viscosity is given by the following expression:

$$\eta = KZ_w^{1.0} \tag{8.11}$$

This relationship represents a simple increase in the viscosity as the polymer chains get longer. For chain lengths above Z_c, we can write the following:

$$\eta = KZ_w^{3.4} \tag{8.12}$$

The dependence on the power of 3.4 arises from the entanglement and diffusion of the polymer chains.

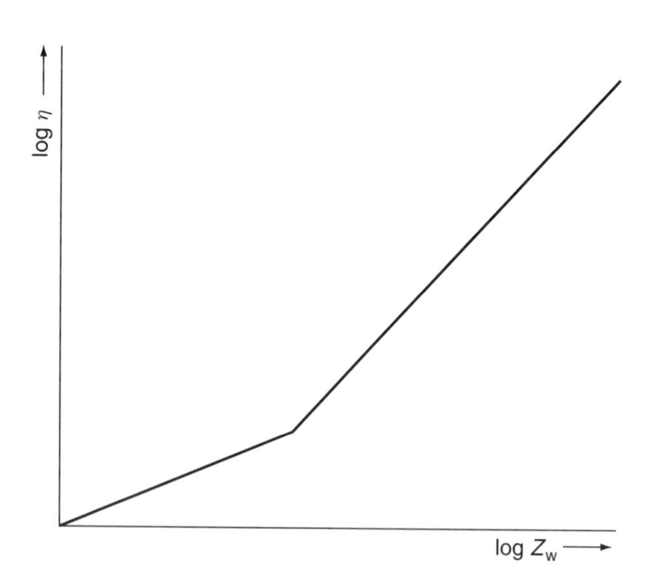

Figure 8.2 Melt viscosity of a polymer as a function of chain length.

SAQ 8.2

A polymer with a weight-average chain length of 200 was found to have a melt viscosity of 100 poise. What is the viscosity of this polymer when the weight average chain length is 800?

The melt viscosity is also temperature-dependent, and the *Williams–Landel–Ferry* (WLF) equation applies as follows:

$$\log(\eta/\eta_{T_0}) = [-C_1(T - T_g)]/[C_2 + (T - T_g)] \quad (8.13)$$

where C_1 and C_2 are constants, and η_{T_0} is the melt viscosity at the T_g or another reference temperature. Generally, η_{T_0} is of the order of 10^{13} poise. If data are not available on the polymer of interest, values of $C_1 = 17.4$ and $C_2 = 51.6$ are usually used. A useful modification of the melt viscosity equation is a combination of equations (8.11) and (8.13), as follows:

$$\log\eta = 1.0\log Z_w + [-C_1(T - T_g)]/[C_2 + (T - T_g)] + k \quad (8.14)$$

This can be applied for polymers with Z_w values below about 600. Equations (8.12) and (8.13) can be combined to give the following:

$$\log\eta = 3.4\log Z_w + [-C_1(T - T_g)]/[C_2 + (T - T_g)] + k \quad (8.15)$$

where k is a constant depending upon the type of polymer. The above equations hold over a temperature range of T_g to (T_g + 100 K).

SAQ 8.3

A polymer with a T_g of 110°C and a weight-average chain length of 400 was found to have a melt viscosity of 5000 poise at 160°C. What is its melt viscosity at 140°C when the weight-average chain length is 900?

8.4 Viscoelasticity

Viscoelasticity is observed when the deformation of the polymer is reversible, but time-dependent [1]. This behaviour is associated with the distortion of polymer chains from their equilibrium conformations through segment motion involving rotation about the chemical bonds in the material.

8.4.1 Creep

Thermoplastics under load display the phenomenon of *creep*, i.e. the deformation continues to increase with time. The amount of strain increases with an increase

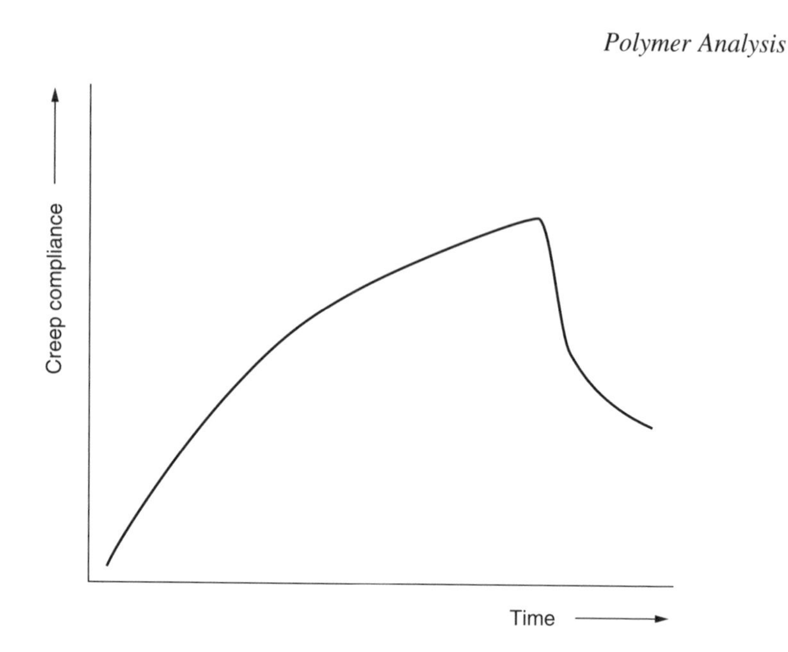

Figure 8.3 A creep curve for a typical viscoelastic plastic material.

in the applied load and temperature. A creep curve for a typical viscoelastic plastic material is shown in Figure 8.3. When the rate of creep is constant, the plastic is subject to viscous flow coupled with cold drawing (*secondary creep*). The viscoelastic effect is shown in plastics when the material is unloaded during secondary creep. Instantaneous elastic recovery is observed, followed by time-dependent recovery (*retarded elasticity*).

Creep is dependent on the molecular weight of a polymer. Above the T_g, amorphous polymers behave like viscous liquids. The creep depends on the molecular weight of the polymer in this case because the creep rate is determined by the polymer viscosity. Below the T_g, the creep of an amorphous polymer is independent of molecular weight as the molecules move by chain segments rather than as whole units. The degree of cross-linking in a polymer also affects the creep in the material. Above the T_g, cross-linking changes the polymer from a viscous liquid to an elastic rubber. Below the T_g, there is very little chain movement and cross-linking has only a small effect unless the degree of cross-linking is very extensive.

The strain after a given time ($\varepsilon(t)$) is related to the applied stress on the sample (σ) by the *creep compliance* ($J(t)$), as follows:

$$J(t) = \varepsilon(t)/\sigma \tag{8.16}$$

The *Boltzmann superposition principle* (BSP) describes the effects of imposing stresses on a polymer at different times, and makes the following statements:

(i) Creep is a function of the entire past loading history of the sample.
(ii) Each loading step makes an independent contribution to the final deformation, which can be obtained by the addition of all of the contributions.

The BSP in its equation form is as follows:

$$\varepsilon(t) = \Sigma \Delta\sigma J(\Delta t) \tag{8.17}$$

This principle is based on two assumptions. First, that the elongation of the polymer sample is proportional to the stress, and secondly, that the elongation due to a given load is independent of the elongation due to any previous load. Thus, the principle is not applicable to crystalline polymers as the first assumption is not valid for such polymers because of the stiffening action of the crystalline phase.

SAQ 8.4

A grade of polypropylene (PP) is found to have the following creep compliance when measured at 35°C:

$$J(t) = 1.2\,t^{0.1}\ \mathrm{GPa}^{-1}$$

where t is expressed in seconds. The polymer is subjected to the following time-sequence of tensile stress at 35°C:

$\sigma = 0 \quad t < 0$

$\sigma = 1$ MPa $\quad 0 < t < 1000$ s

$\sigma = 1.5$ MPa $\quad 1000\ \mathrm{s} < t < 2000$ s

$\sigma = 0 \quad 2000\ \mathrm{s} < t$

Find the strain at the following times:

(a) 1500 s
(b) 2500 s

Assume that under these conditions PP is linearly viscoelastic and therefore obeys the Boltzmann superposition principle.

8.4.2 Models

Viscoelasticity can be represented by using a number of models. A viscoelastic material can be visualized by employing models composed of springs and dashpots. Plastics display elastic behaviour and so we can represent the bond stretching and unstretching from loading and unloading as a spring. In addition, viscous flow can be represented as a dashpot (somewhat like a bicycle pump) with its associated viscosity.

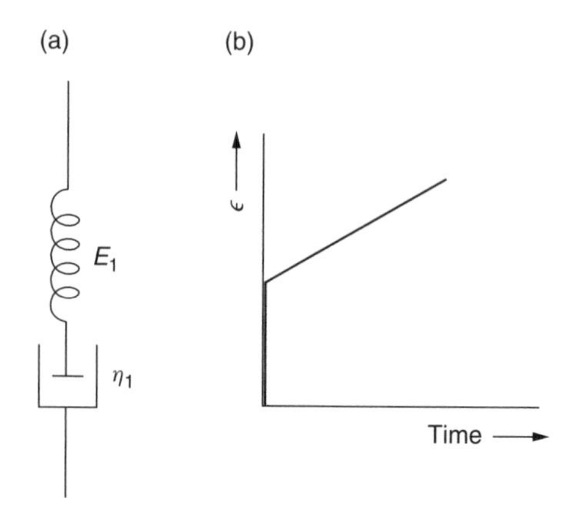

Figure 8.4 Representation of the Maxwell model for viscoelastic behaviour (a) and a typical creep curve obtained for a viscoelastic polymer based on this model (b).

The *Maxwell model* describes a viscoelastic material as a spring in series with a linear dashpot (illustrated in Figure 8.4). The creep curve for a viscoelastic polymer based on the Maxwell model is also shown in Figure 8.4. This curve is approximately consistent with the experimental results, although the unloading curve displays no retarded elasticity. Thus, this model is inadequate as it stands. The creep curve for such a model is given by the following equation:

$$\varepsilon = \sigma/E_1 + \sigma t/\eta \tag{8.18}$$

where σ is the tensile stress, E_1 is the Young's modulus during the instantaneous strain, t is the time, and η is the viscosity.

The model known either as the *Kelvin*, *Voight* or *Kelvin–Voight model* (see Figure 8.5), represents a viscoelastic material as a spring in parallel with a linear

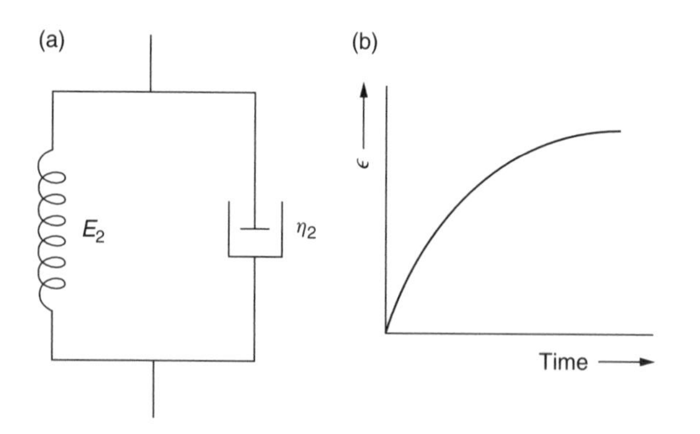

Figure 8.5 Representation of the Kelvin–Voight model for viscoelastic behaviour (a) and a typical creep curve obtained for a viscoelastic polymer based on this model (b).

dashpot. Figure 8.5 also illustrates the creep curve for a viscoelastic polymer based on such a model. The unloading curve has no elastic recovery or residual strain, although retarded elasticity is present. The creep curve in this case is given by the following:

$$\varepsilon = (\sigma/E_2)[1 - \exp(-t/\tau)] \tag{8.19}$$

where E_2 is the Young's modulus during retarded elastic strain, and τ is the retardation time (η/E).

SAQ 8.5

A creep test was carried out with an applied stress of 3 MPa on a polymer specimen. The maximum strain observed was 0.01. After a period of 1 h, the strain was measured to be 0.006. Using the Kelvin–Voight model, calculate the viscosity and modulus of elasticity.

Since the Maxwell and the Kelvin–Voight models between them give creep curves containing all regions of experimental results, a better model for viscoelastic polymers is one which is a combination of elements of both of the models. The *four-element model* (or *combination model*) consists of a combination of series- and parallel-spring dashpot models (Figure 8.6). The creep curve for a viscoelastic polymer based on this combination model is also shown in Figure 8.6. Such a curve can be represented by a combination of the Maxwell and Kelvin–Voight

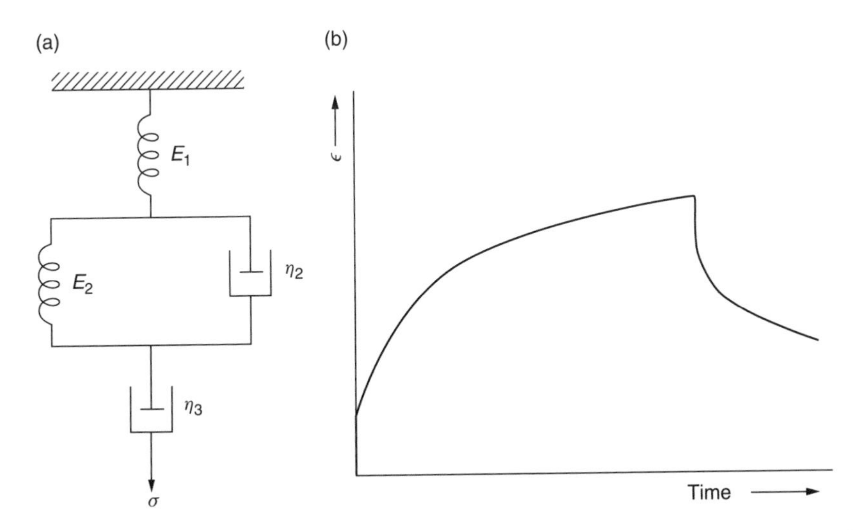

Figure 8.6 Representation of the combination (or four-element) model for viscoelastic behaviour (a) and a typical creep curve obtained for a viscoelastic polymer based on this model (b).

equations, as follows:

$$\varepsilon = \sigma/E_1 + \sigma t/\eta + (\sigma/E_2)[1 - \exp(-t/\tau)] \quad (8.20)$$

Most polymers show deviations and can be better approximated by the following relationship:

$$\varepsilon = \sigma/E + B\sigma^n t + K\sigma[1 - \exp(-qt)] \quad (8.21)$$

where K, B, q and n are experimental constants characteristic of the polymer under investigation.

8.4.3 Stress Relaxation

The stress relaxation of a thermoplastic material under load at constant strain results in a decrease in the stress with time. Such relaxation is caused by the slow sliding of the polymer chains past each other to a more stable random coil state. The rate at which stress relaxation occurs depends on the *relaxation time* (τ), i.e. the time needed for the stress (σ) to decrease to 1/e of the initial stress (σ_0). The decrease in stress with time is given by the following relationship:

$$\sigma = \sigma_0 \exp(-t/\tau) \quad (8.22)$$

This may also be written as follows:

$$\sigma = \sigma_0 \exp(-Et/\eta) \quad (8.23)$$

since $\tau = \eta/E$. The relaxation time is related to the temperature by the following equation:

$$1/\tau = C \exp(-E_A/RT) \quad (8.24)$$

where C is the rate constant (independent of the temperature), E_A is the activation energy for the process, T is the absolute temperature and R is the universal (molar gas) constant.

SAQ 8.6

A polymeric material has a relaxation time of 100 days at 27°C when a stress of 4.0 MPa is applied.

(a) How many days will be required to decrease the stress to 3.2 MPa?
(b) What is the relaxation time at 40°C if the activation energy for this process is 20 kJ mol^{-1}?

In order to characterize the viscoelastic behaviour of amorphous polymers, it is necessary to collect data on relaxations over decades of time, which is obvi-

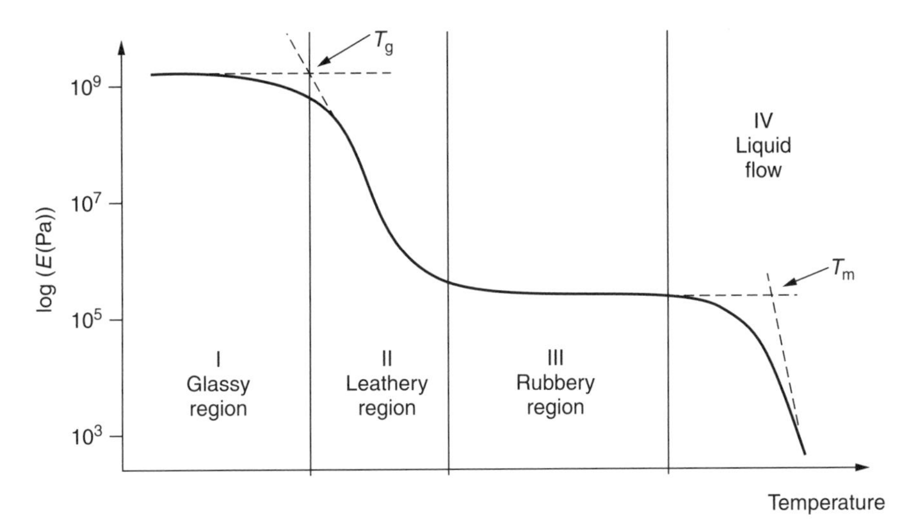

Figure 8.7 Master curve of relaxation modulus as a function of time, used to characterize the viscoelastic behaviour of amorphous polymers.

ously an impractical way to obtain information about the process. This problem can be overcome by exploiting the fact that for most amorphous polymers the deformation for a short period of time at one temperature is equivalent to a longer period at a lower temperature. Thus, it is possible to build up a master curve of the relaxation modulus versus time at a single temperature by the processing of data obtained at a variety of temperatures. Figure 8.7 shows such a typical master curve.

To convert the data obtained at a given experimental temperature (T) to a reference temperature (T_0), a shift factor (a_T) is used, as follows:

$$a_T = \tau_T / \tau_{T_0} \tag{8.25}$$

where τ_T is the relaxation time at temperature T, and τ_{T_0} is the relaxation time at a reference temperature T_0. The value of a_T is obtained from a version of the WLF equation, as follows:

$$\log a_T = [-17.4(T - T_0)]/[51.6 + (T - T_0)] \tag{8.26}$$

The constants in this equation are based on the temperatures being measured in °C. The T_g can be chosen as the reference temperature. It should be noted that the WLF equation holds up to a temperature of 100°C above the T_g. The temperature-dependence of the shift factor is shown in Figure 8.8.

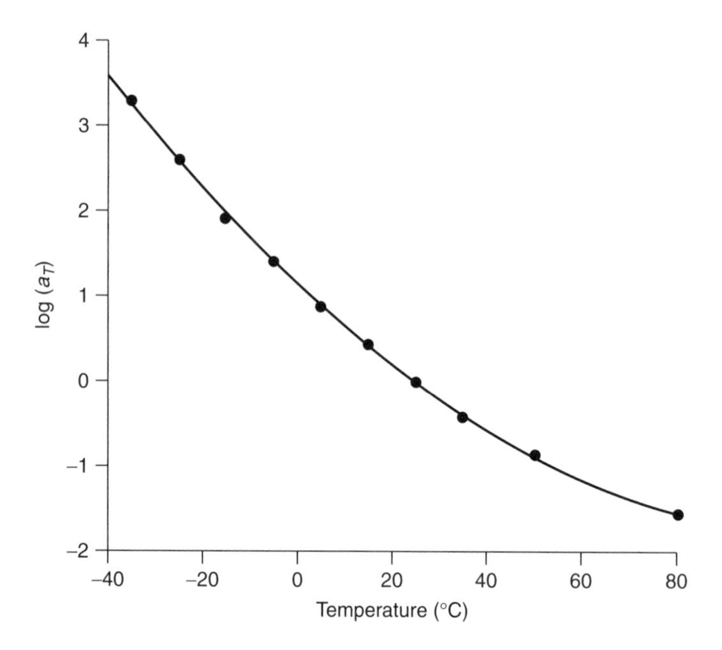

Figure 8.8 Temperature-dependence of the (temperature) shift factor.

SAQ 8.7

(a) Accelerated testing of polyisobutylene at 50°C shows that the relaxation modulus is 10^3 Nm^{-2} at 63 h. Use the shift factor versus temperature curve shown in Figure 8.8 to determine the time for the relaxation modulus to reach 10^3 Nm^{-2} at 25°C.

(b) Use the WLF equation to confirm the above calculation, given that the T_g of polyisobutylene is −70°C.

8.5 Elasticity

In elastomers, the local freedom of the motion associated with small-scale movement of the chain segments is retained, while large-scale movement is prevented by the restraint of the network structure. Most materials under stress exhibit an initial elastic region where the strain is proportional to the stress and if the stress is released the material then returns to its original length. Elastomers sustain reversible strains of the order of hundreds of percent of the original length. A typical stress–strain curve of an elastomer was shown above in Figure 8.1 (curve B). There are some common characteristics shown by elastomers. Such materials are soft and have low elastic moduli, very high strains are possible, the strains are reversible, and the materials are non-crystalline and above the T_g at room temperature.

In an unstressed state, elastomer molecules adopt a randomly coiled conformation. When an elastomer is subjected to stress, the bulk material experiences significant deformation and adopts an extended conformation. When the stress is removed, the molecules revert to their undeformed dimensions. Strain-induced crystallization results when rubber molecules are extended. The effect of such crystallization in natural rubber is illustrated in Figure 8.9. Certain synthetic elastomers, such as styrene–butadiene rubbers, do not undergo this strain-induced crystallization. This elastomer is a random copolymer and hence lacks the molecular regularity necessary to form crystallites on extension.

The kinetic theory of rubber elasticity explains the highly recoverable deformation which is possible in elastomers. This theory is based on the idea that the deformation of an elastomer is analogous to the compression of an ideal gas and a combination of the first and second laws of thermodynamics. When an elastomer chain is extended, the distance between the cross-links in the material increases, thus reducing the number of possible coiled conformations from Ω_0 to Ω. This leads to a decrease in entropy, as follows:

$$S_0 - S = k \ln(\Omega_0/\Omega) \tag{8.27}$$

This equation can be rewritten as follows:

$$S_0 - S = N_0 k[(L/L_0)^2 + 2(L_0/L) - 3]/2 \tag{8.28}$$

where N_0 is the number of chains between cross-links, L is the sample length and L_0 is the original sample length with an original cross-sectional area A. The

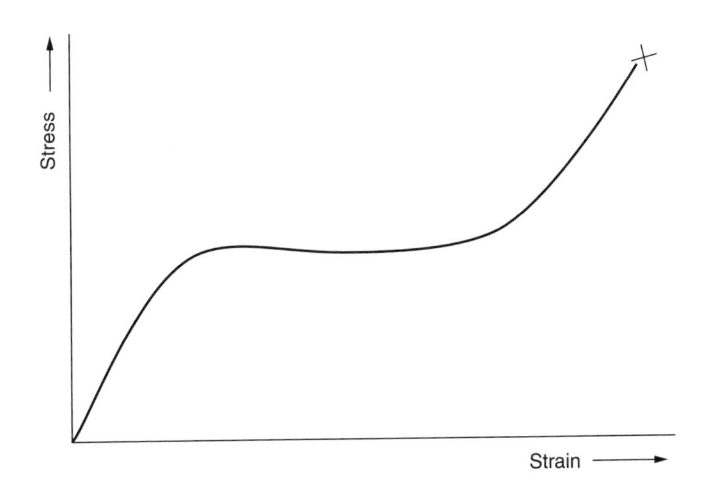

Figure 8.9 The effect of strain-induced crystallization in elastomers.

tensile force for the sample is given by the following:

$$F = N_0kT[(L/L_0) - (L_0/L)^2]/L_0 \quad (8.29)$$

or:

$$F = N_0kT(\lambda - 1/\lambda^2)/L_0 \quad (8.30)$$

where $\lambda(= L/L_0)$ is the extension ratio. Dividing the tensile force by the cross-sectional area gives the stress as follows:

$$\sigma = F/A = nkT[(L/L_0)^2 - L_0/L] = nkT(\lambda^2 - 1/\lambda) \quad (8.31)$$

where n is the number of chain segments between the cross-links per unit volume. Theory agrees well with experiment when the kinetic theory of rubbers is applied, except at very large elongations. In such situations, the molecular chains begin to align and deformation results from the stretching of primary bonds instead of straightening of the originally coiled chain segments.

SAQ 8.8

A bar of rubber, containing 5×10^{20} chains between cross-links, is extended uniaxially at 20°C until its length is double the initial length (0.10 m). What is the force required to double the length of this specimen?

8.6 Processing Methods

There are a number of methods used to process polymers, although processing does involve some common steps [3, 4]. The first of these involves softening the material by heating. Pressure is then applied to the softened polymer to allow it to flow in a die or mould. Finally, the material is allowed to solidify while holding its shape. The choice of a processing method is based on whether the polymer is a thermoplastic or a thermoset, and the shape and number of the final product. When processing thermoplastics, powders or granules are used, while thermosets are processed either from a powder or a liquid monomer. During the processing of thermosets, the cross-linking or curing process is completed.

The most common processing method for thermoplastics is *injection moulding*, and Figure 8.10 illustrates a typical injection moulding machine. The thermoplastic, as a powder or granules, is transferred from a feed hopper and heated in a barrel. The material is next heated by a rotation screw until it becomes softened, and is then forced through a nozzle into a cooler mould. After a set cooling time, the mould is opened and the product is ejected. There are an enormous range of products manufactured by using injection moulding, including computer bodies, compact discs, television housings, crash helmets and telephones. Injection moulding is also used in the manufacture of thermoset products. Given that

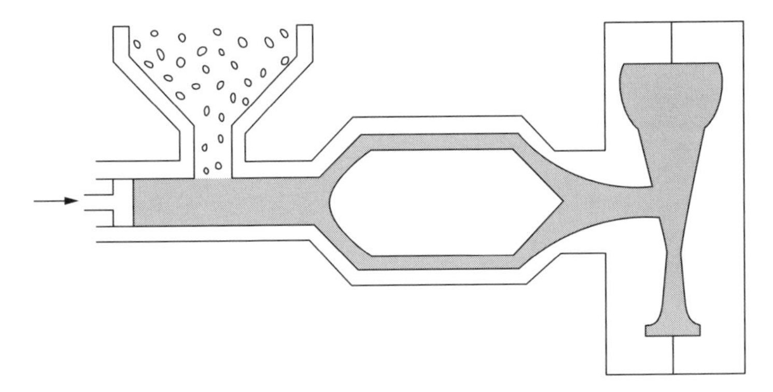

Figure 8.10 Schematic of a typical injection moulding machine.

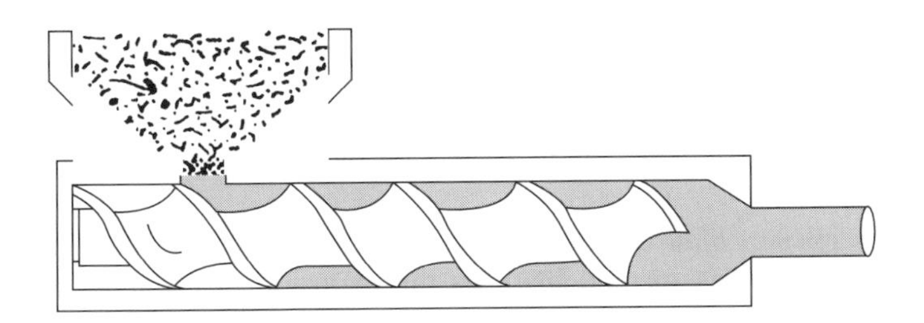

Figure 8.11 Schematic of a typical extrusion apparatus.

thermosets undergo irreversible reaction during processing, specially designed injection barrels and screws are required. In addition, the thermoset is injected into a heated mould and allowed to cure or cross-link before ejecting.

Extrusion is also a common method for the manufacture of thermoplastics. A typical extruding apparatus is illustrated in Figure 8.11. Granules or powder are fed from a hopper to a screw and are then pushed along the barrel chamber to be heated. The polymer is successively compacted, melted and formed into a continuous charge of viscous fluid. The molten material is then forced through a die. The solidification of the extruded material is accelerated by blowers or water sprays just prior to passing on to a conveyor. Extrusion is particularly useful for producing continuous plastic products, such as piping, tubing, rods, sheets and filaments.

Compression moulding is another method used for producing thermosets. This process involves placing partially polymerized thermoset powder into the lower half of a heated mould, while the upper half is then pressed down on top, thus allowing the material to take the shape of the mould. The process of compression moulding is illustrated in Figure 8.12. Heat and pressure accelerate the polymerization. The mould temperatures are usually in the range 130–200°C. Once the cross-linking or curing process is complete, the solid polymer is ejected from the

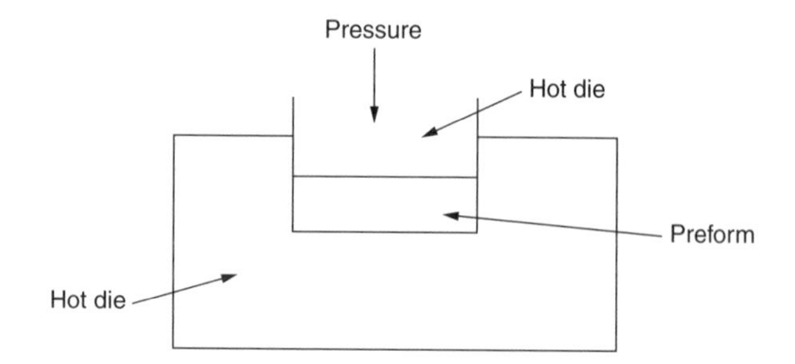

Figure 8.12 Schematic illustrating the process of compression moulding.

mould. Some typical compression moulded products are saucepan handles, lawn bowls and ash trays.

Transfer moulding is also used to manufacture thermoset polymers. In this method, the powder is heated and compressed in a transfer chamber and enters the cavity in a flowing state. Consequently, transfer moulding is usually faster than compression moulding, and is suitable for moulding more intricately shaped samples.

The principles of *blow moulding* grew out of the technique of glass blowing. The first stage of this method involves the extrusion of a length of polymer tubing, known as a *parison*. While this tubing is still in a semi-molten state, it is placed in a mould of the required form. Air is blown into the mould, so forcing the parison against the walls of the mould. The mould is opened when the thermoplastic is cool.

Thermoplastic film and sheeting can be formed by using a *calendering* method. This technique involves squeezing the plastic through a gap (or nip) between two counter-rotating cylinders. Calendering is commonly used for producing PVC sheeting.

Plasticizers are used to improve the flexibility, ductility and toughness of polymers which are intrinsically brittle at room temperature. Plasticizers are usually low-molecular-weight liquids with low vapour pressures. The plasticizer molecules reduce intermolecular bonding by occupying the spaces between the polymer chains. Some common plasticizers include phthalate esters, phosphate esters and adipates. The dominant plasticizer is dioctyl phthalate (DOP), used in PVC.

Fillers are particulate or fibrous solids that are used to improve dimensional stability, compressive strength, abrasion resistance, thermal stability and impact resistance. Fillers help reduce the cost of polymers by replacing the volume of the more expensive (polymer) component. Some common particulate fillers are silica, clay and calcium carbonate, with the latter being the dominant filler in the market. Fibrous fillers include cellulose and polyester fibres.

A variety of pigments or dyes are used as colourants for polymers. Dyes are oil-soluble and dissolve, and are incorporated by mixing with the powdered plastic before processing. Pigments, such as titanium dioxide, are colourants with a small particle size which remain as a separate phase.

8.7 Tensile Testing

Tensile testing machines are designed to elongate samples of materials at a constant rate [3]. Figure 8.13 illustrates the main components of a tensile testing apparatus. The sample is elongated by a moving crosshead. The load cell measures the magnitude of the applied load on the sample, while the extensometer measures the elongation of the sample.

A standard tensile sample of a polymer, in the shape of a 'dog-bone' is illustrated in Figure 8.14. During testing, the deformation occurs to the narrowed central region of the sample which has a uniform cross-sectional area along its length. The sample is mounted by its ends into the grips of the testing apparatus.

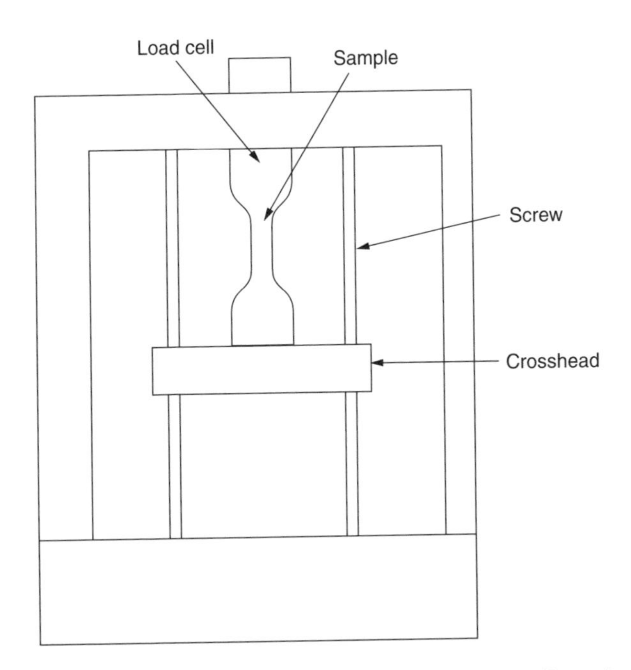

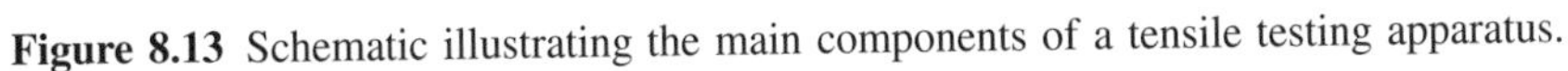

Figure 8.13 Schematic illustrating the main components of a tensile testing apparatus.

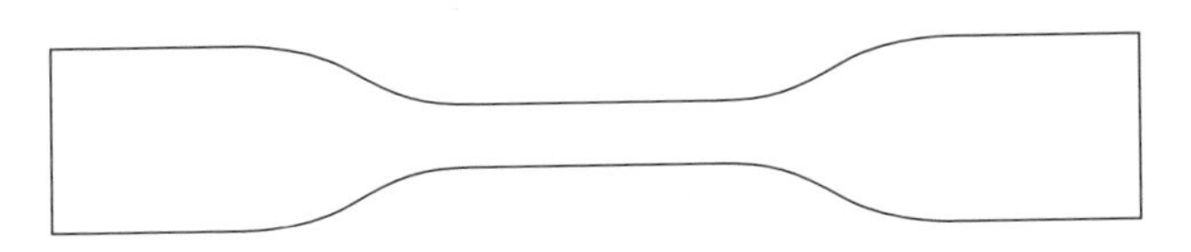

Figure 8.14 A dog-bone tensile test sample.

SAQ 8.9

The force–extension curve for a sample of polystyrene (PS) is shown below in Figure 8.15. The sample is 20 mm wide and 3 mm thick, with a gauge length of 50 mm. Determine the Young's modulus for this PS sample.

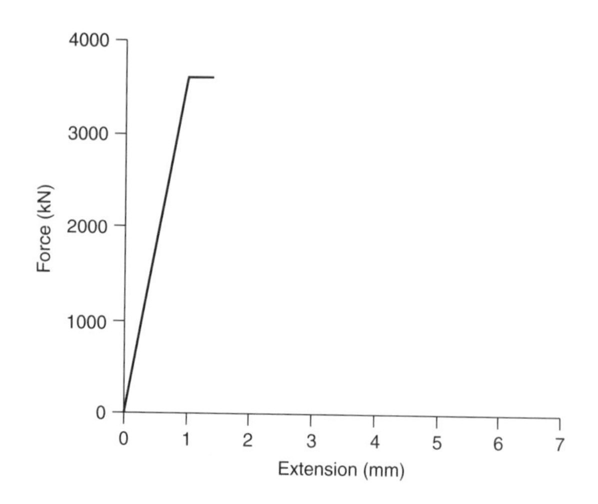

Figure 8.15 The force–extension curve obtained for a sample of polystyrene (SAQ 8.9).

SAQ 8.10

Tensile testing was carried out on a polymer sample with dimensions of 12.6 mm (width), 3.5 mm (thickness) and 50.0 mm (gauge length). Table 8.2 below lists the results of the study. Determine the Young's modulus, the tensile strength and the percentage elongation at break for this polymer. The length between the gauge marks on the test sample at break is 97 mm, and the maximum force in the test was 1290 N. Identify the polymer tested.

Table 8.2 Tensile testing data obtained for a sample of unknown polymer (SAQ 8.9)

Force (N)	Extension (mm)	Force (N)	Extension (mm)
25	0.018	175	0.192
50	0.040	200	0.238
75	0.064	225	0.293
100	0.090	250	0.355
125	0.121	275	0.425
150	0.153	300	0.520

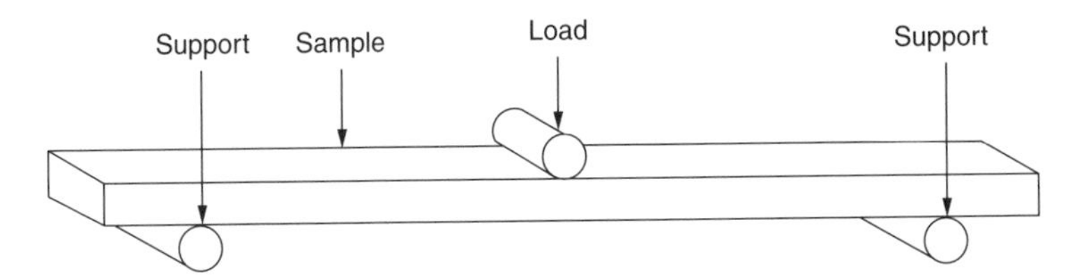

Figure 8.16 Schematic of the apparatus used in the three-point bending test method.

SAQ 8.11

A bar of polypropylene (PP) is of length 200 mm and has a rectangular cross-section of dimensions 25 mm × 3 mm. The sample was subjected to a constant tensile load of 250 N acting along its length. At a time of 100 s after the load was applied, the length is measured and is found to have increased by 0.5 mm. Determine the 100 s creep compliance.

8.8 Flexural Testing

Flexural (or bending) tests are used to measure the rigidity of polymers [3]. In such tests, rectangular bars of the sample to be evaluated are placed on supports located 100 mm apart and a load is applied to the sample at a specified rate. The apparatus used in the *three-point bending test* is illustrated in Figure 8.16. The stress at break is measured and is then used to determine the flexural strength via the following equation:

$$\text{flexural strength} = 3F_{\mathrm{B}}L/2bh^2 \qquad (8.32)$$

where F_{B} is the load at break. Where the stress at break is not measured, the stress at 1.5 times the thickness of the sample is reported. The flexural modulus is given by the following:

$$\text{flexural modulus} = L^3F/4bh^3Y \qquad (8.33)$$

where F/Y is the slope of the initial linear load curve.

8.9 Tear-Strength Testing

The ability to resist tearing is a significant property of certain polymers, such as the thermoplastic films used in packaging [3]. In tear tests, the *tear strength* is the parameter measured and represents the energy required to tear apart a cut specimen with a standard geometry. There are a variety of standard test geometries, including trouser tear and angular tear. The tests can be conducted by using a universal testing machine or specialized testing equipment, such as an Elmendorf machine. The samples used have cuts or slits made before or during the testing procedure.

8.10 Fatigue Testing

Fatigue is a form of failure that can occur when a polymer is subjected to cyclic loading [2]. Such failure occurs at stress levels that are low relative to the yield strength. This property is measured by using a rotating–bending test apparatus, which imposes compression and tensile stresses on the sample as it is simultaneously bent and rotated. The tests are started by subjecting a sample to cyclic stress at a specific maximum stress amplitude and the number of cycles to failure is counted. This procedure is repeated on other samples at progressively decreasing maximum stress amplitudes. The data are plotted as stress (S) versus the logarithm of the number of cycles to failure (N) for each sample (known as a *S–N curve*), as illustrated in Figure 8.17. The higher the magnitude of the stress, then the smaller the number of cycles the polymer is capable of sustaining before failure. For some polymers, the S–N curve becomes horizontal at higher values of N. This is known as the *fatigue limit*, a value below which fatigue failure will not occur. Other polymers do not show a fatigue limit and the S–N curve continues a downward trend with increasing N values. For these polymers, fatigue failure will ultimately occur regardless of the magnitude of the stress. *Fatigue strength* is used to describe the fatigue response in such materials and is defined as the stress level at which failure will occur for some specified number

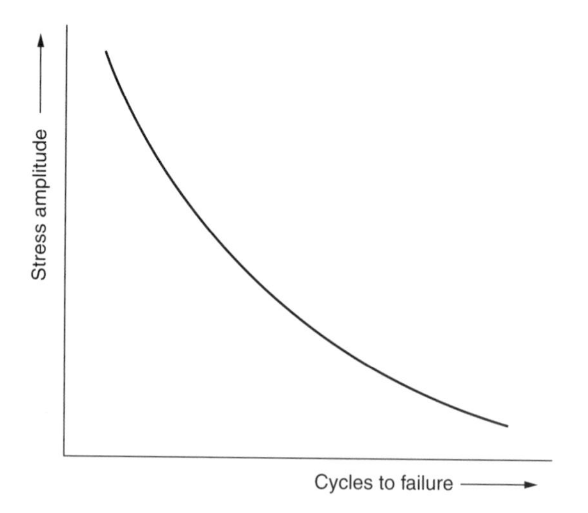

Figure 8.17 A typical S–N curve obtained in the fatigue testing of polymer materials.

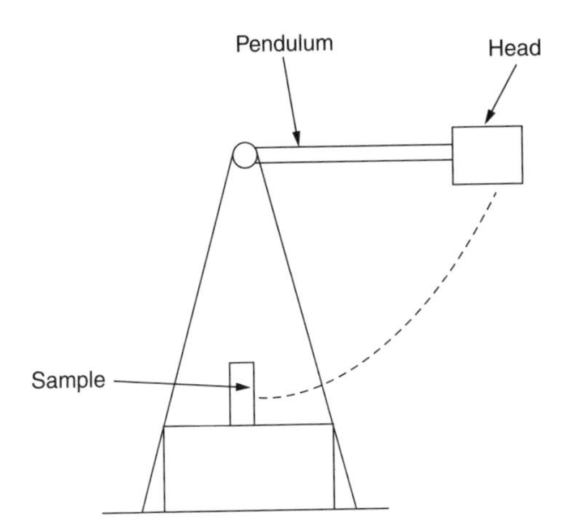

Figure 8.18 Schematic of the set-up used in the impact testing of polymer materials (Charpy or Izod tests).

of cycles. The *fatigue life* is the number of cycles to cause failure at a specified stress level from the $S-N$ curve.

8.11 Impact Testing

Impact testing techniques can be used to understand and evaluate the fracture characteristics of polymers [2]. Two standard impact tests, *Charpy* and *Izod*, are used to measure the impact energy of polymers. In such tests, a sample in the shape of a bar with a V-shaped notch is used. The positioning of the sample varies between the Charpy and Izod tests, with a typical set-up being illustrated in Figure 8.18. The load is applied as a blow from a weighted pendulum hammer released from a fixed height (h). The energy expended in fracture is reflected in the difference between h and the swing height (h'). Impact tests can be used to determine whether a polymer shows a brittle-to-ductile transition with decreasing temperature.

8.12 Hardness Testing

Hardness tests involve the measurement of the resistance of a material to penetration by an indentor [3]. For polymers, several types of hardness tests, involving different shaped indentors, are commonly employed. The *Rockwell hardness* test uses a spherical steel indentor and the hardness is determined from the penetration depth. There are two Rockwell conditions used for polymers, i.e. condition R with a load of 590 N (range R20–R120) and condition M with a load of 980 N (range M20–M140). Generally, rigid thermosets give rise to Rockwell hardness values in the range M100–M120, while engineering thermoplastics tend to have hardnesses in the range R110–R120. The *Shore hardness* test uses pin-shaped indentors. The Shore A test employs a blunt indentor and covers a range of 20A–95A. Shore A is suitable for examining soft polymers such as elastomers. The Shore D test uses a pointed indentor and can be used for examining soft thermoplastics and covers a range of 40D–90D. For more rigid polymers, such as thermosets, the *Barcol hardness* test can be employed. The Barcol indentor is similar to the Shore D indentor, but with a flat tip instead of a round one. The usual range in this case is 50D–90D.

8.13 Viscometry

Capillary rheometers are used to measure the viscosity of undiluted (or melt) polymers [5]. A typical capillary rheometer is illustrated in Figure 8.19. High pressure forces a polymer liquid to pass through a capillary tube, with the flow rate being determined by the plunger speed and the dimensions of the barrel. A capillary viscometer can be used to obtain approximate melt viscosities under various shear stresses by using the equation for melt flow through a capillary. For these viscosity values, the zero-shear viscosity (η_0) can be calculated.

The shear stress at the wall (τ) is given by the following:

$$\tau = LR/2l\pi a^2 \tag{8.32}$$

where L is the load in N, R is the internal radius of the capillary, l is the length of the die and a is the radius of the plunger. The shear rate at the wall (γ) is

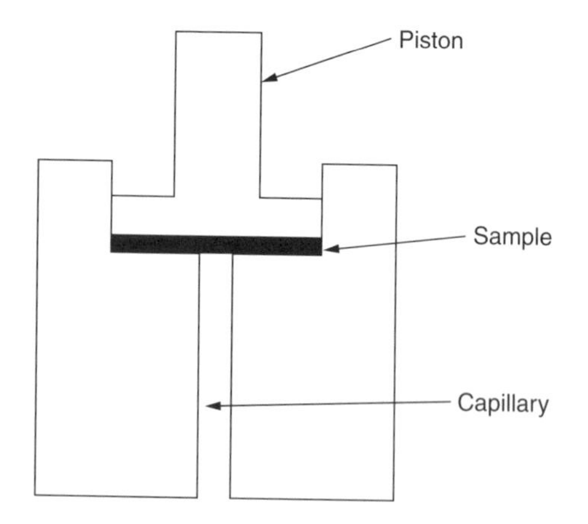

Figure 8.19 Schematic of a typical capillary rheometer used to measure the viscosity of undiluted (or melt) polymers.

given by the following:

$$\dot{\gamma} = 4Q/\pi R^3 \tag{8.33}$$

where Q is the rate of extrusion in mm^3 s^{-1}. Consequently, the viscosity may be determined by using equation (8.9). By plotting $\log \eta$ versus $\log \dot{\gamma}$, the value of the viscosity at zero shear rate can be determined.

SAQ 8.12

A capillary viscometer was used to obtain melt viscosity data for a sample of high-density polyethylene (HDPE) at 190°C, with the data obtained being listed below in Table 8.3. Determine the zero-shear viscosity. The internal radius of the capillary is 1.048 mm, the length of the die is 8.00 mm and the radius of the plunger is 4.735 mm. The density of the PE melt at 190°C is 0.850 g cm^{-3}.

Table 8.3 Melt viscosity data obtained for a sample of high-density polyethylene at 150°C (SAQ 8.12)

Load (g)	Time (s)	Extrudate mass (g)
598	275	0.087
548	230	0.049
498	313	0.054
365	621	0.061
241	1140	0.050

The *melt flow index* (MFI) can also be determined by using a capillary viscometer. The rate of extrusion of a polymer melt is determined through an apparatus of specifically defined dimensions. A heated barrel of diameter 9.57 mm and length 0.80 mm, and with a capillary diameter of 0.209 mm, is used at a specified temperature. The flow is obtained by cutting off the length of extrudate that flows through the capillary in a specified time. As an example, for PE the flow rate measured in g per 10 min is the MFI at a temperature of 190°C and with a piston mass of 2.160 kg. Although measuring the MFI in such a manner has its limitations, it is nevertheless widely used in quality control as it shows very good reproducibility for specific polymers.

8.14 Dynamic Mechanical Analysis

Dynamic tests are useful for evaluating the stress relaxation and the creep behaviour of polymers [6]. If a relaxation process can be described by an Arrhenius relationship, dynamic mechanical analysis (DMA) can be used to determine the activation energy of this process. The slope of a plot of $\ln (f)$ versus $1/T$, where f is the frequency and T is the temperature at which $\tan \delta$ reaches a maximum value, allows the activation energy to be deduced. Figure 8.20 illustrates how DMA may be applied to obtain relaxation data using the WLF equation. First, a series of curves for E' as a function of frequency over a range of temperature is obtained (Figure 8.20 (a)). A master curve is then derived from these data by using the WLF equation to shift the curves along the frequency axis to correspond to the 145°C reference curve (Figure 8.20(b)).

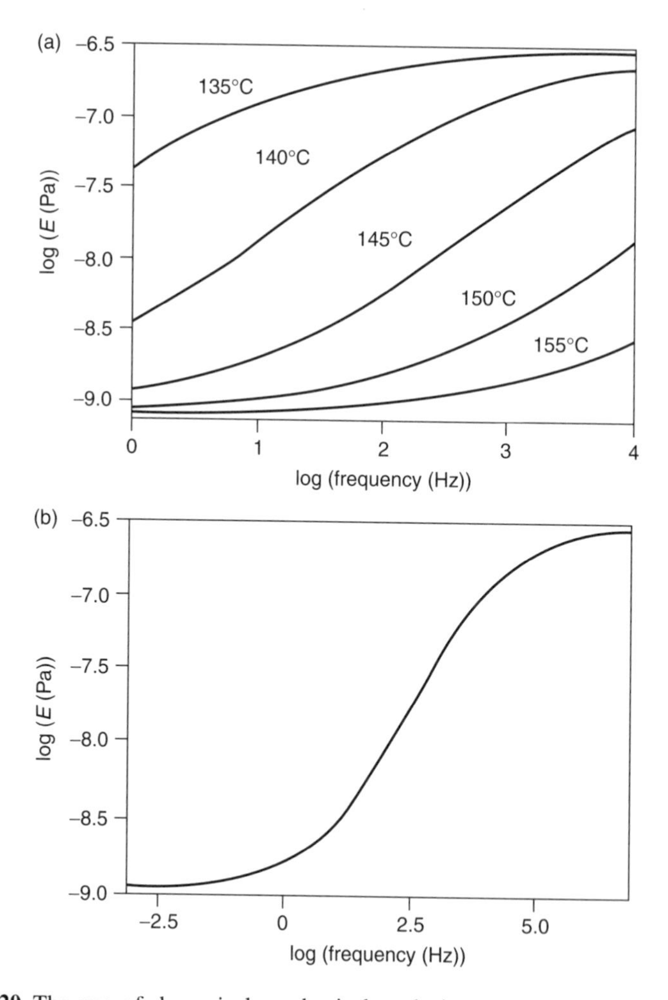

Figure 8.20 The use of dynamical mechanical analysis to evaluate the relaxation behaviour of polymers: (a) plots of E' as a function of frequency over a range of temperatures; (b) the resulting master curve obtained from these data by using the WLF equation. From P. J. Haines, *Thermal Methods of Analysis: Principles, Applications and Problems*, Blackie Academic (1995) Reproduced by permission of Blackie Academic.

Summary

In this chapter, some of the fundamental aspects of the mechanical properties of polymers were introduced. These included stress–strain behaviour, viscous flow, viscoelasticity and elastic behaviour. The processing methods that are commonly used to manufacture polymers were then described. A number of the standard mechanical tests used to characterize polymers were introduced in this chapter, including tensile, flexural, tear strength, fatigue, impact and hardness tests. Other methods that can be used to investigate the flow properties of polymers, including viscometry and dynamic mechanical analysis, were also discussed.

References

1. Ward, I. M. and Hadley, D. W., *An Introduction to the Mechanical Properties of Solid Polymers*, Wiley, Chichester, 1993.
2. Callister, W. D., *Materials Science and Engineering: An Introduction*, 4th Edn, Wiley, New York, 1997.
3. Charrier, J. M., *Polymeric Materials and Processing: Plastics, Elastomers and Composites*, Hanser, Munich, 1991.
4. Michaeli, W., *Plastic Processing: An Introduction*, Hanser, Munich, 1995.
5. Flanagan, M., *Mechanical and Rheological Testing* in *Polymer Characterization*, Hunt B. J. and James M. I. (Eds), Blackie, London, 1993.
6. Haines, P. J., *Thermal Methods of Analysis: Principles, Applications and Problems*, Blackie, London, 1995.

Responses to Self-Assessment Questions

Chapter 1

Response 1.1

The molecular weight (M) of the mer is $(12 \times 2 + 4 \times 1)$ g mer^{-1} = 28 g mer^{-1}. The degree of polymerization is given by the following:

$$\begin{aligned} \text{DP} &= \text{polymer } M/\text{mass of mer} \\ &= 100\,000 \text{ g mol}^{-1}/28 \text{ g mer}^{-1} \\ &= 3.6 \times 10^3 \text{ mer mol}^{-1} \end{aligned}$$

Response 1.2

The polyisoprene structural repeat unit has a molecular weight of 68 g mol^{-1}. Thus:

$$\text{number of moles} = 100 \text{ g}/68 \text{ g mol}^{-1} = 1.5 \text{ mol}$$

For 100% cross-linking, 1.5 mol sulfur is required:

$$\text{mass} = 1.5 \text{ mol} \times 32 \text{ g mol}^{-1} = 48 \text{ g}$$

For 5% cross-linking:

$$5\% \times 48 \text{ g} = 2.4 \text{ g}$$

Response 1.3

The structural repeat unit of polyacrylonitrile (PAN) is $-CH_2-CHCN_n-$, and so the weight of the repeat unit is:

$$(3 \times 12 + 14 + 3 \times 1)\ \text{g} = 53\ \text{g}$$

Likewise, the weight of the structural repeat unit of polystyrene (PS) is:

$$(8 \times 12 + 8 \times 1)\ \text{g} = 104\ \text{g}$$

In order to determine the mole fraction of each component, take the case of 100 g of the copolymer. From this assumption, the number of moles of acrylonitrile in the copolymer is given by:

$$35\ \text{g}/53\ \text{g} = 0.66$$

Likewise, the number of moles of styrene in the copolymer is given by:

$$65\ \text{g}/104\ \text{g} = 0.63$$

The mole fractions are then determined as follows:

$$\text{mole fraction of acrylonitrile} = 0.66/(0.66 + 0.63) = 0.51$$

$$\text{mole fraction of styrene} = 0.63/(0.66 + 0.63) = 0.49$$

Response 1.4

By using equation (1.1):

$$\rho_c = V_m \rho_m + V_f \rho_f$$

$$1515\ \text{kg m}^{-3} = (1 - V_f) \times 1135\ \text{kg m}^{-3} + V_f \times 2500\ \text{kg m}^{-3}$$

and thus:

$$V_f = 0.28$$

Chapter 2

Response 2.1

Consultation of Tables 2.2 and 2.5 allows the number of possibilities to be reduced. The polymer is a thermoplastic as it has a measurable T_m, with a value of 130°C indicating several options, including HDPE, PMMA, PAN and cellulose acetate (see Table 2.5). However, from Table 2.2, the only one of the four polymers that has a density near 1.19 g cm^{-3} is PMMA. The results of the flame and pyrolysis tests support this conclusion (see Tables 2.3 and 2.4). In addition, PMMA is soluble in chloroform and acetone, and insoluble in methanol (see Table 2.1).

Response 2.2

The major infrared modes of PCL observed in Figure 2.3 are listed below in Table 2.13. It is often difficult to differentiate the overlaps which appear in the fingerprint region. The best strategy for spectrum interpretation is to look first at the high-wavenumber end of the spectrum (>1500 cm^{-1}) and concentrate initially on the major bands.

Table 2.13 The major infrared modes of polyacrylonitrile (Response 2.2)

Wavenumber (cm^{-1})	Assignment
2950, 2850	C–H stretching
1720	C=O stretching
1450–900	C–O stretching, C–C stretching
750	CH_2 rocking

Response 2.3

On inspection of the higher-frequency end of the spectrum, bands are observed at 2867 and 2937 cm^{-1}, due to symmetric and asymmetric C–H stretching, respectively. There is also a band at 3300 cm^{-1}, due to N–H stretching, while a strong band at 1640 cm^{-1} is indicative of C=O stretching. The presence of these modes suggests a polyamide and the polymer is, in fact, nylon 6. Hydrogen bonding plays a significant role in the spectra of nylons. The N–H stretching mode is due to hydrogen-bonded groups, while the broad shoulder that appears at around 3450 cm^{-1} is due to non-hydrogen-bonded N–H bonds.

Response 2.4

A plot of the absorbance ratio, 2250 cm^{-1}/1600 cm^{-1}, as a function of concentration is linear and can be used as a calibration plot (see Figure SAQ 2.4 below). The absorbance ratio for the unknown sample is $0.205/0.121 = 1.70$ and consultation of this graph gives a concentration of 26.6% acrylonitrile for the copolymer. Thus, the composition of the copolymer is 73.4% styrene/26.6% acrylonitrile.

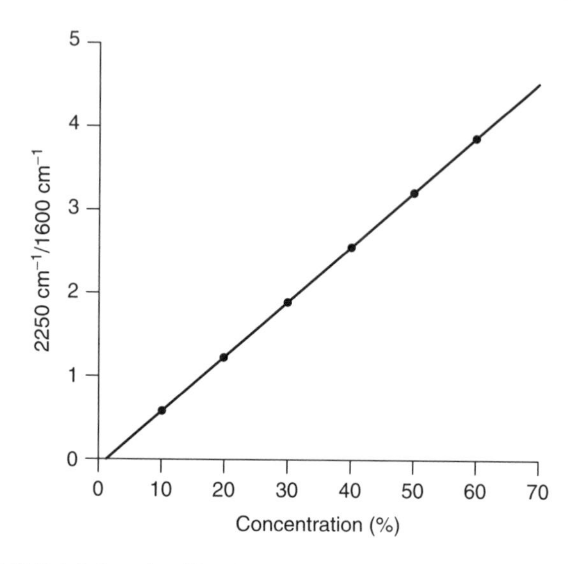

Figure SAQ 2.4 Infrared calibration plot for styrene/acrylonitrile copolymers.

Response 2.5

Table 2.14 below lists the Raman modes of Kevlar. These assignments were made by using the correlation table presented in Figure 2.6.

Table 2.14 The Raman modes of Kevlar (Response 2.5)

Wavenumber (cm^{-1})	Assignment
3080, 3060	Aromatic C–H stretching
1648	C=O stretching, C–N stretching, N–H bending (amide I band)
1610	Ring vibration
1569	N–H bending, C–N stretching (amide II band)
1181, 1277, 1327, 1327	C–C ring stretching

Response 2.6

The ratio of the 1440 cm^{-1} and 1640 cm^{-1} peak heights is 3:1. Consultation of the plot presented in Figure 2.8 shows that there are six methylene groups per repeat unit in this nylon, thus indicating that the Raman spectrum shown in Figure 2.9 is that of nylon 8.

Response 2.7

By using equation (2.4):

$$\% \text{ MMA} = 100 \times (A_{3.6 \text{ ppm}}/3)/(A_{3.6 \text{ ppm}}/3 + A_{3.9 \text{ ppm}}/2)$$
$$= 100 \times (37/3)/(37/3 + 61/2)$$
$$= 29$$

Thus, the copolymer contains 29% MMA and 71% HMA.

Response 2.8

The data presented in Table 2.9 can be used to form a calibration plot as the Beer–Lambert law (equation (2.5)) applies. The plot is shown below in Figure SAQ 2.8 and is linear. An absorbance of 0.349 corresponds to a styrene concentration of 17%. Thus, the composition of the copolymer is 17% styrene/83% butadiene.

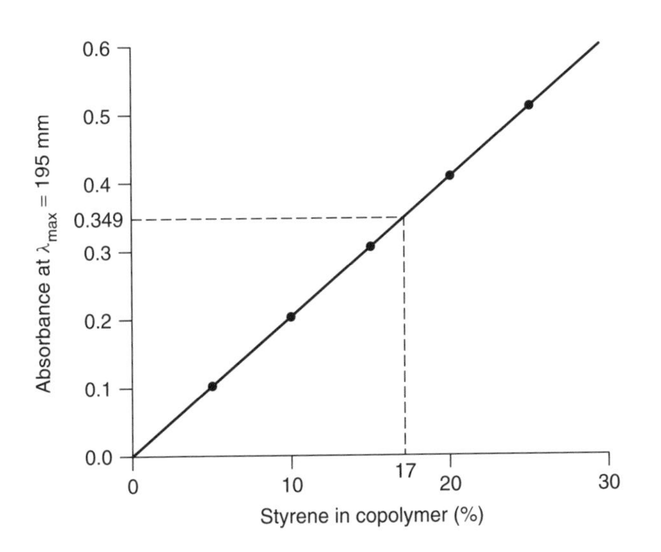

Figure SAQ 2.8 Ultraviolet–visible plot for styrene/butadiene copolymers.

Response 2.9

The peak observed at 130°C can be attributed to HDPE, while the peak at 180°C can be assigned to PP. The enthalpy of the HDPE peak is obtained from its area:

$$1663 \text{ mJ}/10.5 \text{ mg} = 158 \text{ J g}^{-1}$$

while the enthalpy of the PP peak is:

$$126 \text{ mJ}/10.5 \text{ mg} = 12.0 \text{ J g}^{-1}$$

Thus the percentage of PP in the sample is:

$$12.0 \text{ J g}^{-1}/100 \text{ J g}^{-1} \times 100\% = 12.0\%$$

Response 2.10

Using the calibration plot shown in Figure 2.28, a retention volume of 5.5 ml corresponds to an MMA concentration of 36%. Thus, the copolymer contains 64% styrene and 36% methyl methacrylate.

Chapter 3

Response 3.1

Initiation:

$$C_6H_5CO_2CO_2C_6H_5 \longrightarrow 2C_6H_5CO_2^{\bullet}$$

$$C_6H_5CO_2^{\bullet} + CH_2{=}CHCl \longrightarrow C_6H_5CO_2CH_2C^{\bullet}HCl$$

Propagation:

$$C_6H_5CO_2\text{–}CH_2\text{–}CHCl_n\text{–}CH_2C^{\bullet}HCl + CH_2{=}CHCl$$
$$\longrightarrow C_6H_5CO_2\text{–}CH_2\text{–}CHCl_{n+1}\text{–}CH_2C^{\bullet}HCl$$

Termination by combination:

$$C_6H_5CO_2\text{–}CH_2\text{–}CHCl_x\text{–}CH_2C^{\bullet}HCl + C_6H_5CO_2\text{–}CH_2\text{–}CHCl_y\text{–}CH_2C^{\bullet}HCl$$
$$\longrightarrow C_6H_5CO_2\text{–}CH_2\text{–}CHCl_x\text{–}CH_2CHClCHClCH_2\text{–}CH_2\text{–}CHCl_y\text{–}CO_2C_6H_5$$

Response 3.2

(a) As R_p is directly proportional to [M], then simple ratios can be used to determine the propagation rate, as follows:

$$R_{p1}/R_{p2} = [M]_1/[M]_2$$
$$(1.5 \times 10^{-7} \text{ mol l}^{-1} \text{ s}^{-1})/R_{p2} = 1 \text{ M}/2 \text{ M}$$
$$R_{p2} = 3.0 \times 10^{-7} \text{ mol l}^{-1} \text{ s}^{-1}$$

(b) As R_p is proportional to the square root of the initiator concentration, then:

$$R_{p1}/R_{p2} = [I]_1^{1/2}/[I]_2^{1/2}$$
$$(1.5 \times 10^{-7} \text{ mol l}^{-1} \text{ s}^{-1})/R_{p2} = (0.05 \text{ M})^{1/2}/(0.1 \text{ M})^{1/2}$$
$$R_{p2} = 2.1 \times 10^{-7} \text{ mol l}^{-1} \text{ s}^{-1}$$

Response 3.3

Initiation:

$$CH_2{=}CH_2 + H_2SO_4 \longrightarrow CH_3C^+H_2 + HSO_4^-$$

Propagation:

$$\sim CH_2C^+H_2(HSO_4^-) + CH_2{=}CH_2 \longrightarrow \sim CH_2CH_2CH_2C^+H_2(HSO_4^-)$$

Termination – if this is via chain transfer to the counter-ion, then:

$$\sim CH_2C^+H_2(HSO_4^-) \longrightarrow \sim CH_2{=}CH_2 + H_2SO_4$$

while if the termination is through chain transfer to the monomer, then:

$$\sim CH_2C^+H_2(HSO_4^-) + CH_2{=}CH_2 \longrightarrow \sim CH{=}CH_2 + CH_2C^+H_2(HSO_4^-)$$

Response 3.4

Initiation:

$$KNH_2 \longleftrightarrow K^+ + NH_2^-$$

$$NH_2^- + CH_2{=}CH(CN) \longrightarrow NH_2CH_2C^-H(CN)$$

Propagation:

$$NH_2\text{–}CH_2\text{–}CH(CN)_n\text{–}CH_2C^-H(CN) + CH_2{=}CH(CN)$$
$$\longrightarrow NH_2\text{–}CH_2\text{–}CH(CN)_{n+1}\text{–}CH_2C^-H(CN)$$

Termination:

$$NH_2\text{–}CH_2\text{–}CH(CN)_x\text{–}CH_2C^-H(CN) + NH_3$$
$$\longrightarrow NH_2\text{–}CH_2\text{–}CH(CN)_x\text{–}CH_2CH_2(CN) + NH_2^-$$

Response 3.5

As the reaction is second-order, a plot of $1/(1-p)$ versus t will be linear with a slope of c_0k'. The slope of this plot is 2.70 h^{-1}. Thus, the rate constant will be 0.87 l mol^{-1} h^{-1}.

Response 3.6

As it is not known whether this reaction is second- or third-order, a comparison of plots of $1/c$ versus t and $1/c^2$ versus t is required. While the former plot is a curve, the latter plot is linear, and so the reaction is third-order and the reaction is self-catalysed. The slope of this plot is $2k = 2.42 \times 10^{-4}$ l^2 mol^{-2} s^{-1}, and thus the rate constant is $k = 1.21 \times 10^{-4}$ l^2 mol^{-2} s^{-1}. The extent of reaction is given by the following equation:

$$1/(1-p)^2 - 1 = 2c_0^2kt = 2 \times (1.76\ \text{mol l}^{-1})^2 \times 1.21 \times 10^{-4}\ \text{l}^2\ \text{mol}^{-2}\ \text{s}^{-1} \times 86\,400\ \text{s}$$

Thus, $p = 0.88$.

Response 3.7

The number of moles of styrene in the monomer feed is:

$$[\text{M}_1] = 110\ \text{g}/104\ \text{g mol}^{-1} = 1.06\ \text{mol}$$

while the number of moles of vinyl chloride in the feed is:

$$[\text{M}_2] = 200\ \text{g}/62.5\ \text{g mol}^{-1} = 3.20\ \text{mol}$$

Thus, the mole fraction of styrene in the monomer feed is:

$$f_1 = [\text{M}_1]/([\text{M}_1] + [\text{M}_2]) = 1.06/(1.06 + 3.20) = 0.25$$

and it therefore follows that the mole fraction of vinyl chloride in the feed is:

$$f_2 = f_1 - 1 = 0.75$$

From the copolymer equation, the mole fraction of styrene in the copolymer can then be determined as follows:

$$F_1 = (r_1 f_1^2 + f_1 f_2)/(r_1 f_1^2 + 2 f_1 f_2 + r_2 f_2^2)$$
$$= (17 \times 0.25^2 + 0.25 \times 0.75)/(17 \times 0.25^2 + 2 \times 0.75 + 0.02 \times 0.75^2)$$
$$= 0.86$$

Response 3.8

The data given in Table 3.4 can be used to construct a plot of ln $[(h_0 - h_\infty)/(h_t - h_\infty)]$ versus t. This plot is linear and the slope is equal to $k_p(fk_d[\text{I}]/k_t)^{1/2}$. Thus:

$$1.00 \times 10^{-4}\ \text{s}^{-1} = k_p/k_t^{1/2} \times (1 \times 7 \times 10^{-5}\ \text{s}^{-1} \times 1.35 \times 10^{-2}\text{M})^{1/2}$$
$$k_p/k_t^{1/2} = 0.10\ \text{l}^{1/2}\ \text{mol}^{-1/2}\ \text{s}^{-1/2}$$

For the polymerization of methyl methacrylate in benzene, the value of $k_p/k_t^{1/2}$ is 0.14 $\text{l}^{1/2}$ $\text{mol}^{-1/2}$ $\text{s}^{-1/2}$. The agreement between the two values is good, and confirms that the kinetic behaviour of methyl methacrylate is toluene is similar to that observed in benzene.

Response 3.9

The percentage conversion to PVA can be determined by using the following equation:

$$\% \text{ conversion} = (M_0 - M_t)/M_0 \times 100$$

where M_0 is the monomer concentration at initiation and M_t is the monomer concentration at time t. The peak heights in Figure 3.12 were measured and the % conversion then calculated as a function of time. Figure SAQ 3.9 below shows the % conversion plotted against time, from which three phases can be observed in the polymerization process. The first phase between 0 and 10 min is the initiation of the reaction. The rate here is low due to the formation of radicals. The second phase from 10 to 30 min involves the propagation stage, indicated by a linear acceleration of the reaction during this period. Finally, as the viscosity of the solution increases, the reaction rate slows during the 30 to 54 min period.

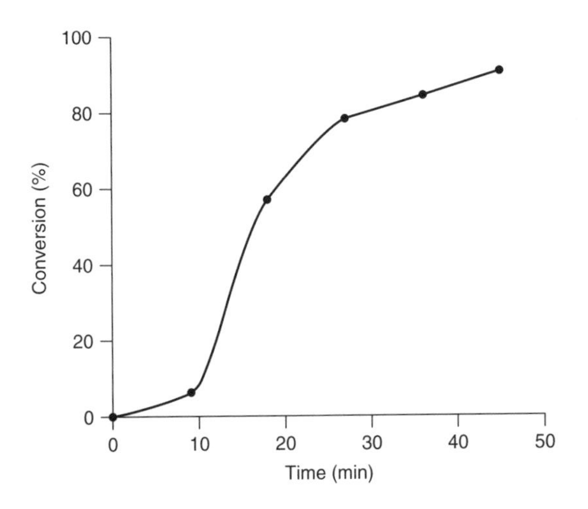

Figure SAQ 3.9 Percentage conversion of vinyl acetate as a function of time.

Chapter 4

Response 4.1

By using equation (4.1):

$$\begin{aligned} \overline{M}_{\rm n} &= \Sigma x_i M_i \\ &= (7500 \times 0.05 + 12\,500 \times 0.16 + 17\,500 \times 0.22 + 22\,500 \\ &\quad \times 0.27 + 27\,500 \times 0.20 + 32\,500 \times 0.08 + 37\,500 \times 0.02)\ \text{g mol}^{-1} \\ &= 21\,200\ \text{g mol}^{-1} \end{aligned}$$

Response 4.2

By using equation (4.2):

$$\overline{M}_{\mathrm{w}} = \Sigma w_i M_i$$
$$= (7500 \times 0.02 + 12\,500 \times 0.10 + 17\,500 \times 0.18 + 22\,500 \times 0.29 + 27\,500 \times 0.26 + 32\,500 \times 0.13 + 37\,500 \times 0.03)\ \mathrm{g\ mol^{-1}}$$
$$= 23\,200\ \mathrm{g\ mol^{-1}}$$

Response 4.3

The structural repeat unit of PS is $-CH_2-CH(C_6H_5)-$, and so $m = 104\ \mathrm{g\ mol^{-1}}$. Thus:

$$n_{\mathrm{n}} = \overline{M}_{\mathrm{n}}/m = 500\,000\ \mathrm{g\ mol^{-1}}/104\ \mathrm{g\ mol^{-1}} = 4800$$

Response 4.4

$$M_0 = 220\ \mathrm{g\ mol^{-1}}$$
$$\overline{M}_{\mathrm{n}} = M_0/(1-p) = 220\ \mathrm{g\ mol^{-1}}/(1-0.96) = 5500\ \mathrm{g\ mol^{-1}}$$
$$\overline{M}_{\mathrm{w}} = M_0(1+p)/(1-p) = 220\ \mathrm{g\ mol^{-1}}(1+0.96)/(1-0.96)$$
$$= 10\,200\ \mathrm{g\ mol^{-1}}$$

Response 4.5

A plot of $(\eta - \eta_0)/\eta_0 c$ versus c shows a y-intercept of $0.0504\ \mathrm{l\ g^{-1}}$, and so $[\eta] = 0.0504\ \mathrm{l\ g^{-1}}$. Rearranging the Mark–Houwink–Sakurada equation then gives:

$$M = ([\eta]/K)^{1/a} = (0.0504\ \mathrm{l\ g^{-1}}/3.80 \times 10^{-8}\ \mathrm{l\ g^{-1}})^{1/0.63} = 90\,500$$

Response 4.6

The molecular weight values for PMMA (M_i) at each of the retention volumes are determined by using the PS calibration information. Solving simultaneous equations gives a expression relating log M_i and V_{R} and this can be used to calculate the M_i values for the PMMA sample – these values are listed below in Table 4.8. The heights obtained in the chromatogram can be converted to weight fractions (w_i) by using equation (4.15). The ($w_i M_i$) and (w_i/M_i) values can then be determined for each fraction. From Table 4.8, the molecular weight values are then calculated as follows:

$$\overline{M}_{\mathrm{n}} = 1/\Sigma(w_i/M_i) = 9570$$
$$\overline{M}_{\mathrm{w}} = \Sigma w_i M_i = 16\,500$$

Table 4.8 SEC data for poly(methyl methacrylate) (Response 4.6)

V_R(ml)	M_i (g mol^{-1})	[h_i (cm)]	$w_i = h_i/\Sigma h_i$	$w_i M_i$	[w_i/M_i]
130	98 000	1.0	3.22×10^{-3}	316	3.30×10^{-8}
135	55 400	12.0	3.86×10^{-2}	2140	6.97×10^{-7}
140	31 300	51.4	0.165	5160	5.27×10^{-6}
145	17 700	89.0	0.286	5060	1.62×10^{-5}
150	9980	84.0	0.270	2690	2.71×10^{-5}
155	5640	51.2	0.165	931	2.93×10^{-5}
160	3190	17.8	0.0573	183	1.80×10^{-5}
165	1800	4.4	0.0142	25.6	7.89×10^{-6}
		$\Sigma h_i = 310.8$		$\Sigma w_i M_i = 16\,500$	$\Sigma(w_i/M_i) = 1.04 \times 10^{-4}$

Response 4.7

A plot of ln (r/r_0) versus t gives a slope of 1.407×10^{-5} $s^{-1} = \omega^2 S$. If 50 000 rpm equals 833 revolutions per second, then:

$$\omega = 2\pi\nu = 2\pi \times 833 \text{ s}^{-1} = 5234 \text{ s}^{-1}$$

The sedimentation constant is therefore:

$$S = 1.407 \times 10^{-5} \text{ s}^{-1}/(5234 \text{ s}^{-1})^2 = 5.136 \times 10^{-13} \text{ s}$$

and thus:

$$\begin{aligned} M &= SRT/(1 - \rho v_s)D \\ &= 5.136 \times 10^{-13} \text{ s} \times 8.134 \text{ J K}^{-1} \text{ mol}^{-1} \times 293 \text{ K}/(1 - 0.9981 \text{ g cm}^{-3} \\ &\quad \times 0.728 \text{ cm}^3 \text{ g}^{-1}) \times 7.62 \times 10^{-7} \times 10^{-4} \text{ m}^2 \text{ s}^{-1} \\ &= 60.1 \text{ kg mol}^{-1} \end{aligned}$$

Response 4.8

As $\pi = \rho$gh, then:

$$h/c = (RT/\rho g\overline{M}_n)(1 + Bc/\overline{M}_n + \cdots \text{etc.})$$

Plotting h/c versus c gives an intercept equal to $RT/\rho g\overline{M}_n$. For such a plot using the data shown in Table 4.6, the intercept is 0.21 cm l g^{-1}. Thus:

$$\begin{aligned} \overline{M}_n &= RT/\rho g(0.21 \text{ cm l g}^{-1}) \\ &= 8.314 \text{ J K}^{-1} \text{ mol}^{-1} \times 298 \text{ K}/980 \text{ kg m}^{-3} \times 9.8 \text{ m s}^{-2} \\ &\quad \times 2.1 \times 10^{-3} \text{ m}^4 \text{ kg}^{-1} \\ &= 1.2 \times 10^2 \text{ kg mol}^{-1} \end{aligned}$$

Response 4.9

Figure SAQ 4.9(a) below shows the $Hc/R(\theta)$ values as a function of concentration. The intercepts obtained from these plots provide $(Hc/R(\theta))_{c=0}$ values which are then plotted against $\sin^2(\theta/2)$, as shown in Figure SAQ 4.9(b). The intercept of this plot equals $1/\overline{M}_w$, so giving a value of $\overline{M}_w$ of 1.30×10^5.

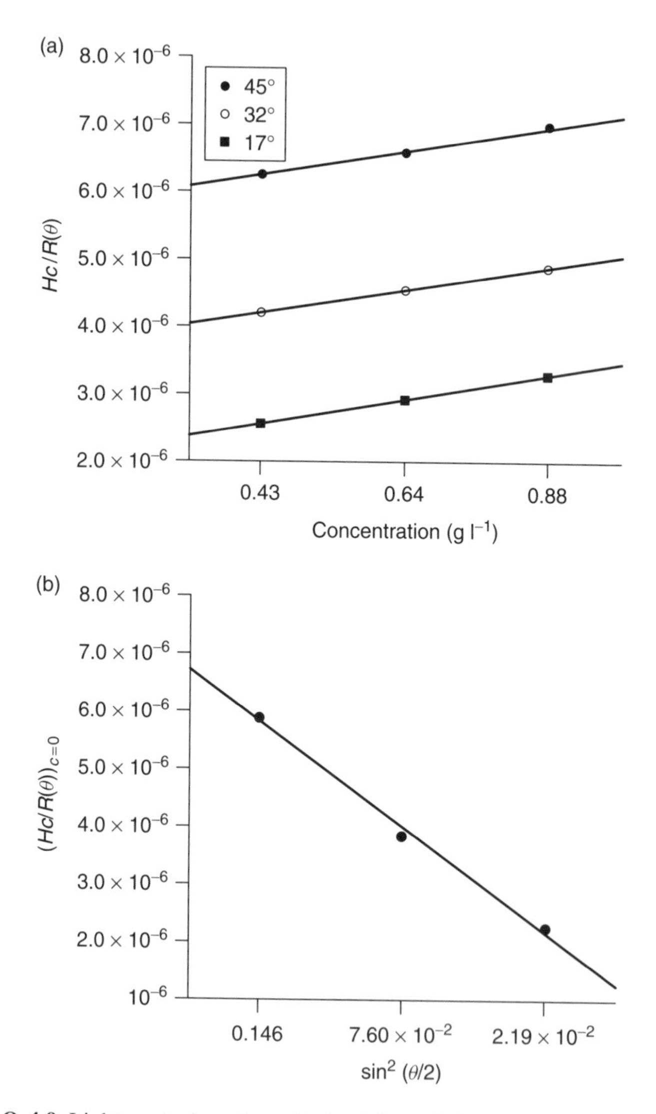

Figure SAQ 4.9 Light scattering plots obtained for solutions of an unknown polymer, using the data given in Table 4.7.

Response 4.10

As this polyamide was prepared with an excess of diamine, the assumption is that the average chain will contain an amine group at each end:

$$H_2N\text{–}(CH_2)_6\text{–}NH\text{–}CO\text{–}(CH_2)_8\text{–}CO_n\text{–}NH_2$$

For this nylon, the amine is titrated and two ends are counted per molecule. The number of moles of H^+ reacting with the nylon is as follows:

$$0.0100 \text{ M} \times 2.450 \times 10^{-2} \text{ l} = 2.45 \times 10^{-4} \text{ mol}$$

Thus, the molecular weight obtained for this nylon sample is:

$$\overline{M}_n = \text{(number of groups which can be determined per polymer molecule)}$$
$$\times \text{(mass of polymer sample/number of moles of H}^+ \text{ reacted with polymer)}$$
$$= 2 \times 2.04 \text{ g}/2.45 \times 10^{-4} \text{ mol}$$
$$= 1.67 \times 10^{-4} \text{ g mol}^{-1}$$

Chapter 5

Response 5.1

$$\text{Number of C–C bonds} = 2 \times 280\,000 \text{ g mol}^{-1}/28 \text{ g mol}^{-1} = 20\,000$$

and so:

$$R_c = 20\,000 \times 154 \times 10^{-12} \text{ pm} = 3.08 \times 10^{-6} \text{ m} = 3.08 \text{ nm}$$

$$R_{rms} = 2N^{1/2}l = 2 \times (20\,000)^{1/2} \times 154 \times 10^{-12} \text{ m} = 3.10 \times 10^{-8} \text{ m} = 31.0 \text{ nm}$$

Response 5.2

By using equation (5.5):

$$\% \text{ crystallinity} = 100 \times \rho_c(\rho_s - \rho_a)/\rho_s(\rho_c - \rho_a)$$

The following simultaneous equations may be written:

$$62.8 = 100 \times \rho_c(0.904 - \rho_a)/0.904(\rho_c - \rho_a)$$

$$54.4 = 100 \times \rho_c(0.895 - \rho_a)/0.895(\rho_c - \rho_a)$$

Solving the above equations gives $\rho_a = 0.842$ and $\rho_c = 0.945 \text{ g cm}^{-3}$.

Thus:

$$74.6 = 100 \times 0.945 \text{ g cm}^{-3}(\rho_s - 0.842 \text{ g cm}^{-3})/\rho_s(0.945 \text{ g cm}^{-3} - 0.842 \text{ g cm}^{-3})$$

and:

$$\rho_s = 0.923 \text{ g cm}^{-3}.$$

Response 5.3

For the first mixture:

$$w_s = 100 \text{ g}/(100 \text{ g} + 25 \text{ g}) = 0.80$$

$$w_l = 1 - 0.80 = 0.20$$

$$1/T_g = w_s/(T_g)_s + w_l/(T_g)_l$$

$$1/273 \text{ K} = 0.80/360 \text{ K} + 0.20/(T_g)_l$$

$$(T_g)_l = 139 \text{ K}$$

Thus, for the second mixture:

$$1/T_g = 0.50/360 \text{ K} + 0.50/139 \text{ K}$$

$$T_g = 200 \text{ K} = -73°\text{C}$$

Response 5.4

$$1/(T_g)_{AB} = w_A/(T_g)_A + w_B/(T_g)_B$$

$$1/(T_g)_{AB} = 0.47/385 \text{ K} + 0.53/195 \text{ K}$$

$$T_g = 254 \text{ K} = -19°\text{C}$$

Response 5.5

By using equation (5.17):

$$1/T_g = w_1/T_{g1} + w_2/T_{g2}$$

$$= 0.5/698 \text{ K} + 0.5/423 \text{ K}$$

$$T_g = 527 \text{ K} = 254°\text{C}$$

Response 5.6

The Avrami plot is given by $\log_{10}\{-\ln[(h_t - h_\infty/(h_0 - h_\infty)]\}$ versus $\log_{10} t$. The plot for this particular system is shown below in Figure SAQ 5.6. The slope of this plot is 2.5, and thus the Avrami exponent for the crystallization of this elastomer at -5°C is $n = 2.5$. The Avrami exponent is not an integer in this case, but lies in the range which indicates one- and two-dimensional growth geometry.

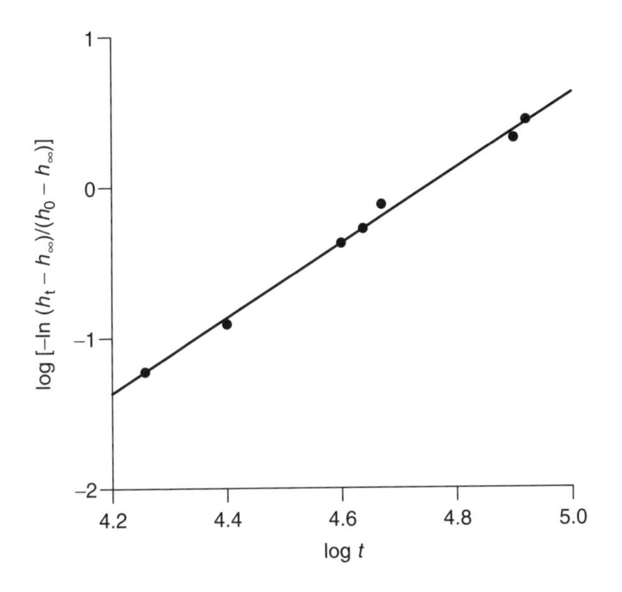

Figure SAQ 5.6 Avrami plot obtained for the crystallization of an unknown elastomer at -5°C, using the data given in Table 5.3.

Response 5.7

In order to demonstrate that the PBI/PC blend in miscible, it must be established whether or not the blend shows a single T_g obeying the Fox equation. Figure 5.23 shows that below a temperature of around 250°C, the percentage of carbonyl absorbance of the PC component remains constant. Above 250°C, the absorbance declines sharply with increasing temperature. This change corresponds to the T_g determined from the Fox equation (equation (5.17)):

$$1/T_g = w_1/T_{g1} + w_2/T_{g2}$$
$$= 0.5/698 \text{ K} + 0.5/423 \text{ K}$$
$$T_g = 527 \text{ K} = 254^\circ\text{C}$$

The change is associated with a thermal disruption of the hydrogen bonding between the N–H groups of PBI with the C=O groups of the PC.

Response 5.8

There are a number of notable differences between the spectra shown in Figure 5.28. In (b), there are multiple peaks in the range 1.5–2.5 ppm. According to Table 5.5, absorptions in this range are due to methylene CH_2. In (a), the main peak in this range is near 2 ppm. These observations suggest that (a) represents predominantly syndiotactic PMMA, while (b) represents isotactic PMMA. This conclusion is supported by the observation of the differences in position of the backbone methyl peak. Figure 5.28(a) shows an intense peak at 1.1 ppm due to syndiotactic PMMA, while Figure 5.28(b) shows an intense peak at 1.3 ppm due to isotactic PMMA.

Response 5.9

Looking first at the data for the PEEK/LCP blend, it can be seen that the T_g of the PEEK is not notably affected by the presence of the LCP in the blend. The Fox equation predicts that the T_g would be concentration-dependent if the blend was miscible, and so this blend is immiscible. Secondly, the data given in Table 5.6 show that the T_g of PEEK in the PEEK/PEI blend is dependent on the concentration of the polyimide in the blend. Figure SAQ 5.9 below presents the T_g values for the PEEK/PEI blend predicted by the Fox equation, as well as the experimental values for this blend, as a function of composition. This figure demonstrates that the predicted and experimental values are similar, and thus the PEEK/PEI blend is miscible over this concentration range. (The data obtained for the PEEK/LCP system are also shown in the figure for comparison purposes.)

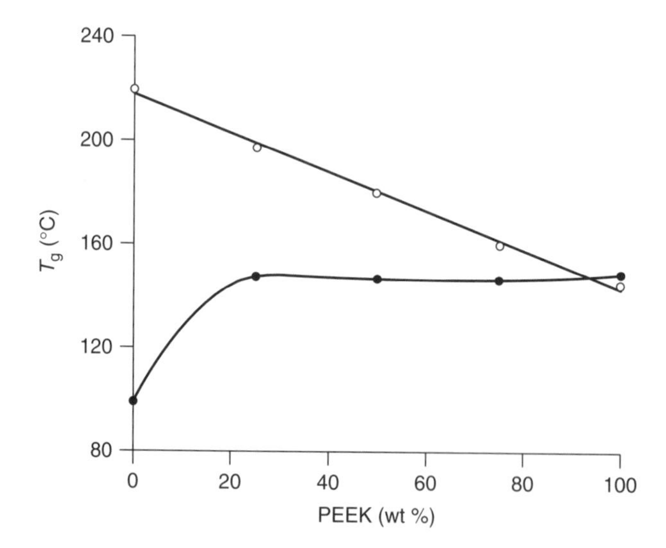

Figure SAQ 5.9 Values of the T_g as a function of composition for the PEEK/PEI and PEEK/LCP blend systems; ○, PEEK/PEI; •, PEEK/LCP.

Response 5.10

The structural repeat unit of PE is $-CH_2-CH_{2n}-$, and so the molecular weight of this unit is 28 g mol^{-1}. For the crystalline sample, the number of moles is:

$$n = m/M = 9.6 \times 10^{-3} \text{ g}/28 \text{ g mol}^{-1} = 3.4 \times 10^{-4} \text{ moles per unit}$$

The enthalpy for the pure crystalline standard is:

$$\Delta H_c = 2.65 \text{ J}/3.4 \times 10^{-4} \text{ mol} = 7.8 \times 10^3 \text{ J mol}^{-1} = 7.8 \text{ kJ mol}^{-1}$$

The number of moles of the commercial PE is:

$$n = m/M = 6.8 \times 10^{-3} \text{ g}/28 \text{ g mol}^{-1} = 2.4 \times 10^{-4} \text{ moles per unit}$$

Thus, the enthalpy change for the unknown sample is:

$$\Delta H = 1.10 \text{ J}/2.4 \times 10^{-4} \text{ mol} = 4.6 \times 10^3 \text{ J mol}^{-1} = 4.6 \text{ kJ mol}^{-1}$$

Hence, the percentage crystallinity of the commercial sample is:

$$100\Delta H/\Delta H_c = 100 \times 4.6 \text{ kJ mol}^{-1}/7.8 \text{ kJ mol}^{-1} = 59\%$$

Response 5.11

From the plot shown in Figure 5.34, $\log E' = 9.3$, and so $E' = 2.0 \times 10^9$ Pa (at 135°C). Thus:

$$\begin{aligned}\tan\delta &= E'/E'' \\ &= 450 \times 10^6 \text{ Pa}/2.0 \times 10^9 \text{ Pa} = 0.23\end{aligned}$$

Response 5.12

From the slopes of the plots of diameter versus time at both temperatures, the growth rates of the spherulite at 135 and 140°C are 0.28 and 0.13 μm s^{-1}, respectively. Equation (5.6) can be rewritten as follows:

$$\ln(\nu/\nu_0)T = (-E_D/R) - [CT_m/(T_m - T)]$$

Thus, substitution of the values at the two temperatures allows the above equation to be solved, giving:

$$E_D = 55 \text{ kJ mol}^{-1} \text{ and } C = 151 \text{ K}$$

Response 5.13

From equation (5.32):

$$\begin{aligned}\tan\theta &= x/R \\ &= 22.9 \text{ mm}/30 \text{ mm}\end{aligned}$$

and therefore:

$$\theta = 37.4^\circ$$

Thus:

$$d = \lambda/2\sin\theta$$
$$= 1.54\ \text{Å}/2 \times \sin 37.4^\circ = 1.27\ \text{Å}$$

Chapter 6

Response 6.1

(a) $\lambda = 1/1000\ \text{cm}^{-1} = 10^{-5}\ \text{m}$

$$d_p = (\lambda/n_1)/\{2\pi[\sin\theta - (n_1/n_2)^2]^{1/2}\}$$
$$= (10^{-5}\ \text{m}/1.5)\{2\pi[\sin 60^\circ - (1.5/2.4)^2]^{1/2}\}$$
$$= 1.5\ \mu\text{m}$$

In a similar way, we obtain the following:
(b) 1.0 μm
(c) 0.5 μm

Note the difference in the depth of penetration at 3000 cm^{-1} when compared to that at 1000 cm^{-1}.

Response 6.2

The data for the 0.003 and 2 μm film samples can be used to obtain a relationship between the carbonyl absorbance and the film thickness of PMMA, as follows:

$$\log A = -0.61 + 0.63 \log x$$

Substitution of the absorbance of the unknown into the above equation then gives a film thickness of 0.07 μm.

Response 6.3

Comparison of Figures 6.7 and 6.9 shows that peaks due to C–C, C–O and O–C=O are present both before and after treatment. However, Figure 6.9 shows two additional components at 285.7 and 286.9 eV, which can be attributed to C–N and C=N functionalities, respectively. The C–C and C=O components have also been affected by the amine treatment, with the relative intensities of both peaks increasing twofold as a result of such treatment.

Response 6.4

The slope of the linear region below the T_g in Figure 6.10 is 6.4×10^3 K^{-1}. Thus, the enthalpy shown below T_g is 6.4×10^3 K^{-1} $\times$ 8.314 J K^{-1} mol^{-1} = 53 kJ mol^{-1}.

Likewise, the slope of the linear region above T_g is 4.2×10^3 K^{-1}, and so the enthalpy in this case is 4.2×10^3 K^{-1} $\times$ 8.314 J K^{-1} mol^{-1} = 35 kJ mol^{-1}.

Chapter 7

Response 7.1

From Table 7.1, the solubility parameter for *n*-hexane is $\delta_1 = 14.8$ J$^{1/2}$ cm$^{-3/2}$, while the same parameter for PE is $\delta_2 = 16.2$ J$^{1/2}$ cm$^{-3/2}$, Thus:

$$|\delta_1 - \delta_2| = |-1.4 \text{ J}^{1/2} \text{ cm}^{-3/2}|$$

This value is less than 3.4 J$^{1/2}$ cm$^{-3/2}$, and so PE is soluble in *n*-hexane.

Response 7.2

The TGA data show that the PP is the least thermally stable of the polymers studied. The PP curve is to the left of the other curves, thus indicating that PP decomposes at temperatures well below the other polymers.

Response 7.3

Figure 7.7 shows that there are two distinct steps in the degradation process. The first degradation step between 200 and 350°C involves a loss of 64% of the sample mass. The maximum rate of mass loss in this step occurs at 300°C. There is also a second degradation step in the range 400–500°C. There is a mass loss of 28% associated with this step, with the maximum rate of mass loss occurring at 450°C. Above 500°C, there is a final product which is stable up to 800°C, with 8% of the original sample remaining at the end of the experiment.

Chapter 8

Response 8.1

Young's modulus is obtained by using equation (8.5) and the data presented in Table 8.1:

$$\begin{aligned} E_c &= E_f V_f + E_m V_m \\ &= 70 \text{ GPa} \times 0.45 + 2.8 \text{ GPa} \times (1 - 0.45) \\ &= 33 \text{ GPa} \end{aligned}$$

The tensile strength is determined from equation (8.6):

$$\sigma_c = \sigma_f V_f + \sigma_m V_m$$
$$= 1800 \text{ MPa} \times 0.45 + 65 \text{ MPa} \times (1 - 0.45)$$
$$= 846 \text{ MPa}$$

Response 8.2

Equation (8.11) applies for Z_w below 600, and thus K can be determined as follows:

$$\eta = K Z_w^{1.0}$$
$$100 \text{ poise} = \text{K} \times 200^{1.0}$$

and so:

$$K = 0.5 \text{ poise}$$

Equation (8.12) can then be applied for a Z_w of 800:

$$\eta = K Z_w^{3.4}$$
$$\eta = 0.5 \text{ poise} \times 800^{3.4}$$
$$\eta = 3.7 \times 10^9 \text{ poise}$$

Response 8.3

From equation (8.14), the constant k for this system may be determined as follows:

$$\log \eta = 1.0 \log Z_w + [-C_1(T - T_0)]/[C_2 + (T - T_0)] + k$$
$$\log(5000) = 1.0 \log (400) + [-17.4(160 - 110)/(51.6 + 160 - 110)] + k$$

and so:

$$k = 9.7$$

For Z_w above 600, equation (8.15) may be used:

$$\log \eta = 3.4 \log (900) + [-17.4(140 - 110)/(51.6 + 140 - 110)] + 9.7$$
$$\log \eta = 13.3$$
$$\eta = 1.9 \times 10^{13} \text{ poise}$$

Response 8.4

(a) $\varepsilon(t) = \Sigma \Delta\sigma J(t)$

$$\varepsilon(1500 \text{ s}) = (1 - 0) \times 10^6 \times 1.2 \times 10^{-9} \times (1.0 \times 10^3 - 0)^{0.1}$$
$$+ (1.5 - 1) \times 10^6 \times 1.2 \times 10^{-9} \times (1.5 \times 10^3 - 10^3)^{0.1}$$
$$= 3.5 \times 10^{-3}$$

(b) $\varepsilon(2500 \text{ s}) = (1 - 0) \times 10^6 \times 1.2 \times 10^{-9} \times (1.0 \times 10^3 - 0)^{0.1}$

$$+ (1.5 - 1) \times 10^6 \times 1.2 \times 10^{-9} \times (2.0 \times 10^3 - 10^3)^{0.1}$$
$$+ (0 - 1.5) \times 10^6 \times 1.2 \times 10^{-9} \times (2.5 \times 10^3 - 2 \times 10^3)^{0.1}$$
$$= 2.4 \times 10^{-4}$$

Response 8.5

The modulus of elasticity is given by the following:

$$E = \sigma/\varepsilon_{\text{max}} = 3 \text{ MPa}/0.01 = 300 \text{ MPa}$$

From the Kelvin–Voight model, we have:

$$\varepsilon = (\sigma/E_2)[(1 - \exp(-t/\tau)]$$
$$0.006 = (3 \text{ MPa}/300 \text{ MPa})[1 - \exp(-1 \text{ h/t})]$$

and so:

$$\tau = 1.1 \text{ h}$$

Thus:

$$\varepsilon = E\tau = 300 \text{ MPa} \times 1.1 \text{ h} = 333 \text{ MPa h} = 1.18 \times 10^{12} \text{ Pa s}$$

Response 8.6

(a) $\ln(\sigma/\sigma_0) = -t/\tau$. Therefore:

$$\ln(3.2 \text{ MPa}/4.0 \text{ MPa}) = -t/100 \text{ days}$$

and so:

$$t = 222 \text{ days}$$

(b) $1/t = C\exp(-E_A/RT)$. Therefore:

$$1/100 \text{ days} = C\exp(-20 \times 10^3 \text{ J mol}^{-1}/8.314 \text{ JK}^{-1} \text{ mol}^{-1} \times 300 \text{ K})$$
$$C = 30 \text{ days}^{-1}$$
$$1/\tau = 30 \text{ days}^{-1} \times \exp(-20 \times 10^3 \text{ J mol}^{-1}/8.314 \text{ J K}^{-1} \text{ mol}^{-1}$$
$$\times 313 \text{ K})$$

and so:

$$\tau = 71 \text{ days}$$

Response 8.7

(a) From the plot shown in Figure 8.8, $\log a_{25°C} = 0$, and so $a_{25°C} = 1$, while $\log a_{50°C} = -0.9$, and so $a_{50°C} = 0.13$. Thus:

$$a_{25°C}/a_{50°C} = 1/0.13 = 7.7$$

and:

$$\tau_{25°C} = 7.7 \times 63 \text{ h} = 485 \text{ h}$$

(b) From the WLF equation:

$$\log a_T = [-17.4(T - T_0)]/[51.6 + (T - T_0)]$$

$$\log a_{25°C} = [-17.4(25 + 70)]/[51.6 + (25 + 70)] = -11.5$$

$$\log a_{50°C} = [-17.4(50 + 70)]/[51.6 + (50 + 70)] = -12.2$$

Thus:

$$\log (a_{25°C}/a_{50°C}) = -11.5/-12.2 = 0.94$$

and:

$$a_{25°C}/a_{50°C} = 8.8$$

which gives:

$$\tau a_{25°C}/a_{50°C} = 8.8 \times 63 \text{ h} = 554 \text{ h}$$

The difference between the values determined for (a) and (b) is most likely to be due to (inaccurate) readings from the plot of shift factor versus temperature (Figure 8.8).

Response 8.8

By using equation (8.30):

$$\begin{aligned} F &= N_0 kT(\lambda - 1/\lambda^2)/L_0 \\ &= 5 \times 10^{20} \times 1.38 \times 10^{-23} \text{ J K}^{-1} \times 293 \text{ K}(2 - 1/2^2)/0.10 \text{ m} \\ &= 35 \text{ N} \end{aligned}$$

Response 8.9

This sample is brittle, as indicated by the linear stress–strain curve, and so the slope can be used to determine E. The slope obtained from this curve is 3.6×10^6

N mm^{-1}. Young's modulus is then calculated by using equation (8.3), as follows:

$$\begin{aligned} E &= (\text{slope} \times \text{gauge length})/(\text{cross-sectional area}) \\ &= 3.6 \times 10^6 \text{ N mm}^{-1} \times 50 \text{ mm}/20 \text{ mm} \times 3 \text{ mm} \\ &= 3.0 \times 10^3 \text{ N mm}^{-2} \\ &= 3.0 \text{ GPa} \end{aligned}$$

Response 8.10

The force–extension data are plotted below in Figure SAQ 8.10. There is no linear elastic region on this graph, and so Young's modulus is determined by using the 0.2% secant modulus. As 0.2% of 50 mm is 0.1 mm, the force for an extension of 0.1 mm is 112 N. Thus, Young's modulus is given by the following:

$$\begin{aligned} E &= (\text{slope} \times \text{gauge length})/(\text{cross-sectional area}) \\ &= (112 \text{ N} \times 50 \text{ mm})/(0.1 \text{ mm} \times 12.6 \text{ mm} \times 3.5 \text{ mm}) \\ &= 1.27 \times 10^3 \text{ N mm}^{-2} \\ &= 1.27 \text{ GPa} \end{aligned}$$

The tensile strength is given by:

$$\begin{aligned} \text{UTS} &= F/A \\ &= 1290 \text{ N}/12.6 \text{ mm} \times 3.5 \text{ mm} \\ &= 29.3 \text{ MPa} \end{aligned}$$

The percentage elongation is:

$$\begin{aligned} &(\text{increase in gauge length} \times 100)/(\text{original gauge length}) \\ &= (97 \text{ mm} - 50 \text{ mm}) \times 100/50 \text{ mm} \\ &= 94\% \end{aligned}$$

The shape of the curve indicates a soft thermoplastic. On comparison with Table 8.1, the tensile data determined here indicate that the polymer is most likely to be HDPE or PP.

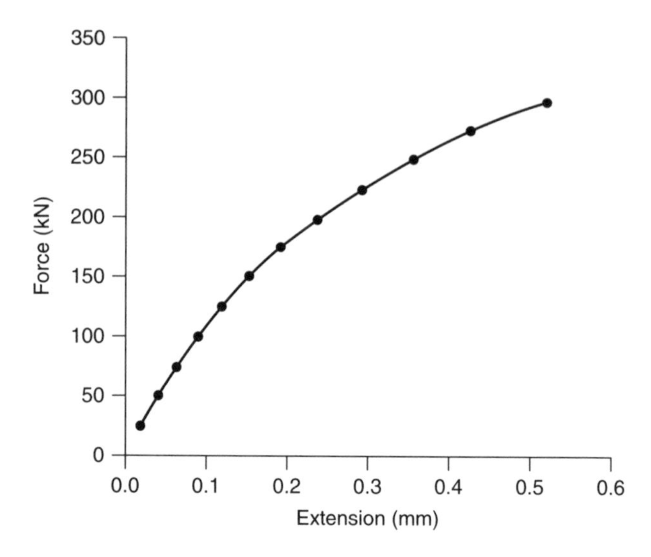

Figure SAQ 8.10 Force–extension plot obtained for an unknown polymer sample, using the tensile testing data given in Table 8.2.

Response 8.11

The applied stress on the sample can be determined by using equation (8.1), as follows:

$$\sigma = F/A = 250\ \text{N}/25 \times 10^{-3}\ \text{m} \times 3 \times 10^{-3}\ \text{m} = 3.3 \times 10^{6}\ \text{Pa}$$

The strain after 100 s is calculated by using equation (8.2):

$$\varepsilon(100\ \text{s}) = \Delta l/l = 0.5 \times 10^{-3}\ \text{m}/200 \times 10^{-3}\ \text{m} = 2.5 \times 10^{-3}$$

Thus, from equation (8.16), we obtain:

$$J(100\ \text{s}) = \varepsilon(100\ \text{s})/\sigma = 2.5 \times 10^{-3}/3.3 \times 10^{6}\ \text{Pa} = 7.5 \times 10^{-10}\ \text{Pa}^{-1}$$

Response 8.12

Table 8.4 below summarizes the results obtained from analysis of the melt viscosity data. The shear stress at the wall is calculated by using equation (8.32), while the flow rate in g s^{-1} can be used to determine the rate of extrusion, i.e. flow rate/density. Thus, the shear rate at the wall is calculated by using equation (8.33). These parameters can then be used to determine the required viscosity by employing equation (8.9). In this case, the intercept of a plot of $\log \eta$ versus $\log \gamma$ gives a value of η_0 of 1.0×10^{4} Pa s^{-1}.

Table 8.4 Melt viscosity data obtained for a sample of HDPE at 90°C (Response 8.12)

Load (N)	Shear stress (N m^{-2})	Flow rate (g s^{-1})	Shear rate (s^{-1})	Viscosity (Pa s^{-1})
5.9	5.3×10^3	0.37	0.41	1.3×10^4
5.4	5.0×10^3	0.25	0.28	1.8×10^4
4.9	4.6×10^3	0.20	0.22	2.1×10^4
3.6	3.3×10^3	0.12	0.13	2.5×10^4
2.4	2.2×10^3	0.052	0.058	3.8×10^4

Bibliography

General

Allen, G. (Ed.), *Comprehensive Polymer Science*, Pergamon Press, Oxford, UK, 1989.

Billmeyer, F. W., *Textbook of Polymer Science*, Wiley, New York, 1984.

Brandrup, J., Immergut, E. H. and Grulke, E. A. (Eds), *Polymer Handbook*, 4th Edn, Wiley, New York, 1999.

Campbell, I. M., *Introduction to Synthetic Polymers*, Oxford University Press, Oxford, UK, 2000.

Kroschwitz, J. I. (Ed.), *Encyclopedia of Polymer Science and Engineering*, 1st Edn, Wiley, New York, 1987.

Kroschwitz, J. I. (Ed.), *Concise Encyclopedia of Polymer Science and Engineering*, 2nd Edn, Wiley, New York, 1998.

Nicholson, J. W., *The Chemistry of Polymers*, Royal Society of Chemistry, Cambridge, UK, 1991.

Painter, P. C. and Coleman, M. M., *Fundamentals of Polymer Science: An Introductory Text*, Technomic Publishing, Lancaster, PA, 1997.

Rudin, A., *The Elements of Polymer Science and Engineering*, Academic Press, San Diego, CA, 1999.

Stevens, M. P., *Polymer Chemistry: An Introduction*, Oxford University Press, Oxford, UK, 1999.

Walton, D. J. and Lorimer, J. P., *Polymers*, Oxford University Press, Oxford, UK, 2000.

Young, R. J. and Lovell, P. A., *Introduction to Polymers*, Chapman and Hall, London, 1991.

Techniques

Beamson, G. and Briggs, D., *High Energy XPS of Organic Polymers: The Scienta ESCA 300 Database*, Wiley, Chichester, 1992.

Bovey, F. A., *High Resolution NMR of Macromolecules*, Academic Press, New York, 1972.

Bower, D. I. and Maddams, W. F., *The Vibrational Spectroscopy of Polymers*, Cambridge University Press, Cambridge, UK, 1989.

Braun, D., *Simple Methods for Identification of Plastics*, Hanser, Munich, 1996.

Crompton, T. K., *Analysis of Polymers: An Introduction*, Pergamon Press, Oxford, UK, 1989.

Fawcett, A. H., *Polymer Spectroscopy*, Wiley, Chichester, 1996.

Haines, P. J., *Thermal Methods of Analysis: Principles, Applications and Problems*, Blackie, London, 1995.

Hendra, P., Jones, C. and Warnes, G., *Fourier Transform Raman Spectroscopy: Instrumentation and Chemical Applications*, Ellis Horwood, New York, 1991.

Higgins, J. S. and Benoit, H. C., *Polymers and Neutron Scattering*, Clarendon Press, Oxford, UK, 1996.

Hunt, B. J. and James M. I. (Eds), *Polymer Characterization*, Blackie, London, 1993.

Koenig, J. L., *Spectroscopy of Polymers*, Elsevier, Amsterdam, The Netherlands, 1999.

Komoroski, R. A. (Ed.), *High Resolution NMR Spectroscopy of Synthetic Polymers in Bulk*, VCH, Deerfield Beach, FL, 1986.

Liebman, S. A. and Levy, E. V. (Eds), *Pyrolysis and GC in Polymer Analysis*, Marcel Dekker, New York, 1985.

Pethrick, R. A. and Dawkins, J. V. (Eds), *Modern Techniques for Polymer Characterization*, Wiley, Chichester, 1999.

Rabek, J. F., *Experimental Methods in Polymer Chemistry*, Wiley, Chichester, 1980.

Ranby, B. and Rabek, J. F., *ESR Spectroscopy in Polymer Research*, Springer-Verlag, Berlin, 1977.

Sandler, S. R., Karo, W., Bonesteel, J. and Pearce, E. M., *Polymer Synthesis and Characterization: A Laboratory Manual*, Academic Press, San Diego, CA, 1998.

Sawyer, L. C. and Grubb, D. T., *Polymer Microscopy*, Chapman and Hall, London, 1987.

Schroder, E., Muller, G. and Arndt, K. F., *Polymer Characterization*, Hanser Publishers, Munich, 1989.

Siesler, H. W. and Holland-Moritz, K., *Infrared and Raman Spectroscopy of Polymers*, Marcel Dekker, New York, 1980.

Spells, S. J. (Ed.), *Characterization of Solid Polymers: New Techniques and Developments*, Chapman and Hall, London, 1994.

Glossary of Terms

This section contains a glossary of terms, all of which are used in the text. It is not intended to be exhaustive, but to explain briefly those terms which often cause difficulties or may be confusing to the inexperienced reader.

Anisotropy The observation of different physical properties in different directions.

Autoacceleration A phenomenon observed for concentrated polymer solutions where the polymerization shows a marked deviation from first-order kinetics.

Birefringence An optical phenomenon observed when polymer chains are aligned in the process of orientation.

Blend A mixture of polymeric materials consisting of at least two polymers or copolymers.

Cage effect A phenomenon observed where free radicals can recombine before they have time to move apart.

Chain transfer A phenomenon observed in free-radical polymerization where the reactivity of a radical may be transferred to another species.

Chemical shift The difference in the absorption position of a nuclear magnetic resonance peak from that of a reference proton.

Chromophore A functional group responsible for electronic absorption.

Cold-drawing A phenomenon observed for thermoplastics subjected to continuous stress above the yield point, manifested as a 'neck' in the sample.

Colourant A dye or pigment added to a polymer to impart a specific colour to the material.

Compatibilization A process of modification of the interfacial properties of an immiscible polymer blend to improve the adhesion and blend properties.

Composite A material composed of a mixture of two or more components or phases.

Contour length The distance from the beginning to the end of the molecule along the covalent bonds of the backbone.

Copolymer A material consisting of polymer chains containing two or more different types of monomers.

Coupling constant The frequency difference between a doublet peak in a nuclear magnetic resonance spectrum proportional to the effectiveness of the coupling.

Creep A form of deformation which continues to increase with time.

Degree of polymerization The number of mers in a polymer structure.

Elastomer A cross-linked polymer showing rubber-like characteristics.

Filler An additive used to improve the mechanical properties of polymers.

Flame retardant A polymer additive used to minimize flammability.

Fluorescence Emission occurring from the lowest excited single state to the singlet ground state.

Fourier transformation A mathematical method involving the interconversion of the two domains of distance and frequency.

Friedel–Crafts catalysts Lewis acids used to catalyse cationic polymerizations.

Fringed-micelle model The model that predicts that polymer crystallization occurs by bundles coming together in regular segments from different molecules.

Group frequencies Infrared modes that may be assigned to particular parts of a molecule.

High-performance polymer A polymer that can withstand high-temperature engineering environments.

Inhibitor A retarder that is very effective and allows no polymer to be initially formed.

Interpenetrating polymer network An assembly of at least two polymers in network form, one of which is prepared or cross-linked in the presence of the other.

Intrinsically conducting polymer A polymer with conjugated π-electrons in the backbone which can be doped to produce a material that exhibits an electrical conductivity approaching that of a metal.

Lamellae model The model that predicts that polymer crystallization takes place by single molecules folding themselves at intervals of about 10 nm to form lamellae.

Limiting-oxygen index A value used to quantify the combustion behaviour of polymers, determined by combusting a polymer sample of a specific size with varying oxygen–nitrogen mixtures.

Liquid crystalline polymer A polymer containing a liquid crystalline state, which is neither strictly crystalline nor liquid.

Living polymer An unterminated polymer produced during anionic polymerization which may be stored and further reacted at a later stage.

Lyotropic The term used to describe liquid crystalline polymers which exhibit liquid crystalline behaviour in solution.

Mer The repeat unit in a polymer.

Newtonian liquid A liquid for which the viscosity is independent of the shear rate.

Phosphorescence Emission occurring from the lowest excited triplet state to the singlet ground state.

Plasticizer An additive used to improved the flexibility of polymers.

Polydispersity index A measure of the breadth of the molecular weight distribution of a polymer.

Radius of gyration The root-mean-square distance of the elements of a chain from its centre of gravity.

Raman scattering The scattering resulting from the interaction of electromagnetic radiation with a molecule, leading to a change in the frequency.

Retarder A substance that can react with a radical to form product(s) incapable of adding monomer.

Root-mean-square end-to-end distance The average distance between the first and last segments of a polymer molecule.

Spin–spin coupling A phenomenon observed in nuclear magnetic resonance spectroscopy resulting from the indirect coupling of nuclei through the intervening bonding electrons

Stabilizer A polymer additive used to counteract degradation.

Tacticity The orientation of a monomer adding to a growing asymmetric polymer chain.

Thermoplastic A polymer which melts when heated and resolidifies when cooled.

Thermoplastic elastomer A polymer which at ambient temperatures exhibits elastomeric behaviour while remaining fundamentally thermoplastic in character.

Thermoset A polymer which does not melt when heated, but decomposes irreversibly at high temperatures.

Thermotropic The term used to describe liquid crystalline polymers which exhibit liquid crystalline behaviour in the melt.

Viscoelasticity A phenomenon observed for a polymer where the deformation of the material is reversible, but time-dependent.

Vulcanization The cross-linking process, involving sulfur, used for natural rubber.

Ziegler–Natta catalysts The complexes between main-group metal alkyls and transition-metal salts used in coordination polymerizations.

SI Units and Physical Constants

SI Units

The SI system of units is generally used throughout this book. It should be noted, however, that according to present practice, there are some exceptions to this, for example wavenumber (cm^{-1}) and ionization energy (eV).

Base SI units and physical quantities

Quantity	Symbol	SI unit	Symbol
length	l	metre	m
mass	m	kilogram	kg
time	t	second	s
electric current	I	ampere	A
thermodynamic temperature	T	kelvin	K
amount of substance	n	mole	mol
luminous intensity	I_v	candela	cd

Prefixes used for SI units

Factor	Prefix	Symbol
10^{21}	zetta	Z
10^{18}	exa	E
10^{15}	peta	P
10^{12}	tera	T
10^{9}	giga	G
10^{6}	mega	M
10^{3}	kilo	k

(continued overleaf)

Prefixes used for SI units *(continued)*

Factor	Prefix	Symbol
10^2	hecto	h
10	deca	da
10^{-1}	deci	d
10^{-2}	centi	c
10^{-3}	milli	m
10^{-6}	micro	μ
10^{-9}	nano	n
10^{-12}	pico	p
10^{-15}	femto	f
10^{-18}	atto	a
10^{-21}	zepto	z

Derived SI units with special names and symbols

Physical quantity	SI unit		Expression in terms of base or derived SI units
	Name	Symbol	
frequency	hertz	Hz	$1\ \mathrm{Hz} = 1\mathrm{s}^{-1}$
force	newton	N	$1\ \mathrm{N} = 1\ \mathrm{kg\ m\ s}^{-2}$
pressure; stress	pascal	Pa	$1\ \mathrm{Pa} = 1\ \mathrm{N\ m}^{-2}$
energy; work; quantity of heat	joule	J	$1\ \mathrm{J} = 1\ \mathrm{N\ m}$
power	watt	W	$1\ \mathrm{W} = 1\ \mathrm{J\ s}^{-1}$
electric charge; quantity of electricity	coulomb	C	$1\ \mathrm{C} = 1\ \mathrm{A\ s}$
electric potential; potential difference; electromotive force; tension	volt	V	$1\ \mathrm{V} = 1\ \mathrm{J\ C}^{-1}$
electric capacitance	farad	F	$1\ \mathrm{F} = 1\ \mathrm{C\ V}^{-1}$
electric resistance	ohm	Ω	$1\Omega = 1\ \mathrm{V\ A}^{-1}$
electric conductance	siemens	S	$1\ \mathrm{S} = 1\Omega^{-1}$
magnetic flux; flux of magnetic induction	weber	Wb	$1\ \mathrm{Wb} = 1\ \mathrm{V\ s}$
magnetic flux density; magnetic induction	tesla	T	$1\ \mathrm{T} = 1\ \mathrm{Wb\ m}^{-2}$
inductance	henry	H	$1\ \mathrm{H} = 1\ \mathrm{Wb\ A}^{-1}$
Celsius temperature	degree Celsius	°C	$1°\mathrm{C} = 1\ \mathrm{K}$

Derived SI units with special names and symbols *(continued)*

Physical quantity	SI unit		Expression in terms of base or derived SI units
	Name	Symbol	
luminous flux	lumen	lm	1 lm = 1 cd sr
illuminance	lux	lx	$1 \text{ lx} = 1 \text{ lm m}^{-2}$
activity (of a radionuclide)	becquerel	Bq	$1 \text{ Bq} = 1 \text{ s}^{-1}$
absorbed dose; specific energy	gray	Gy	$1 \text{ Gy} = 1 \text{ J kg}^{-1}$
dose equivalent	sievert	Sv	$1 \text{ Sv} = 1 \text{ J kg}^{-1}$
plane angle	radian	rad	1[a]
solid angle	steradian	sr	1[a]

[a] rad and sr may be included or omitted in expressions for the derived units.

Physical Constants

Recommended values of selected physical constants[a]

Constant	Symbol	Value
acceleration of free fall (acceleration due to gravity)	g_n	$9.806\,65 \text{ m s}^{-2}$[b]
atomic mass constant (unified atomic mass unit)	m_u	$1.660\,540\,2(10) \times 10^{-27}$ kg
Avogadro constant	L, N_A	$6.022\,136\,7(36) \times 10^{23} \text{ mol}^{-1}$
Boltzmann constant	k_B	$1.380\,658(12) \times 10^{-23} \text{ J K}^{-1}$
electron specific charge (charge-to-mass ratio)	$-e/m_e$	$-1.758\,819 \times 10^{11} \text{C kg}^{-1}$
electron charge (elementary charge)	e	$1.602\,177\,33(49) \times 10^{-19}$C
Faraday constant	F	$9.648\,530\,9(29) \times 10^{4} \text{C mol}^{-1}$
ice-point temperature	T_{ice}	273.15 K[b]
molar gas constant	R	$8.314\,510(70) \text{ J K}^{-1} \text{ mol}^{-1}$
molar volume of ideal gas (at 273.15 K and 101 325 Pa)	V_m	$22.414\,10(19) \times 10^{-3} \text{ m}^3 \text{ mol}^{-1}$
Planck constant	h	$6.626\,075\,5(40) \times 10^{-34}$ J s
standard atmosphere	atm	101 325 Pa[b]
speed of light in vacuum	c	$2.997\,924\,58 \times 10^{8} \text{ m s}^{-1}$[b]

[a] Data are presented in their full precision, although often no more than the first four or five significant digits are used; figures in parentheses represent the standard deviation uncertainty in the least significant digits.

[b] Exactly defined values.

The Periodic Table

0.98 — Pauling electronegativity
3 — Atomic number
Li — Element
6.941 — Atomic weight (^{12}C)

Group : 3 ← d transition elements → 12

Group 1	Group 2	3	4	5	6	7	8	9	10	11	12	Group 13	Group 14	Group 15	Group 16	Group 17	Group 18
1 2.20 **H** 1.008																	2 **He** 4.003
3 0.98 **Li** 6.941	4 1.57 **Be** 9.012											5 2.04 **B** 10.811	6 2.55 **C** 12.011	7 3.04 **N** 14.007	8 3.44 **O** 15.999	9 3.98 **F** 18.998	10 **Ne** 20.179
11 0.93 **Na** 22.990	12 1.31 **Mg** 24.305											13 1.61 **Al** 26.98	14 1.90 **Si** 28.086	15 2.19 **P** 30.974	16 2.58 **S** 32.064	17 3.16 **Cl** 35.453	18 **Ar** 39.948
19 0.82 **K** 39.102	20 1.00 **Ca** 40.08	21 **Sc** 44.956	22 **Ti** 47.90	23 **V** 50.941	24 **Cr** 51.996	25 **Mn** 54.938	26 **Fe** 55.847	27 **Co** 58.933	28 **Ni** 58.71	29 **Cu** 63.546	30 **Zn** 65.37	31 1.81 **Ga** 69.72	32 2.01 **Ge** 72.59	33 2.18 **As** 74.922	34 2.55 **Se** 78.96	35 2.96 **Br** 79.909	36 **Kr** 83.80
37 0.82 **Rb** 85.47	38 0.95 **Sr** 87.62	39 **Y** 88.906	40 **Zr** 91.22	41 **Nb** 92.906	42 **Mo** 95.94	43 **Tc** (99)	44 **Ru** 101.07	45 **Rh** 102.91	46 **Pd** 106.4	47 **Ag** 107.87	48 **Cd** 112.40	49 1.78 **In** 114.82	50 1.96 **Sn** 118.69	51 2.05 **Sb** 121.75	52 2.10 **Te** 127.60	53 2.66 **I** 126.90	54 **Xe** 131.30
55 0.79 **Cs** 132.91	56 0.89 **Ba** 137.34	57 **La** 138.91	72 **Hf** 178.49	73 **Ta** 180.95	74 **W** 183.85	75 **Re** 186.2	76 **Os** 190.2	77 **Ir** 192.22	78 **Pt** 195.09	79 **Au** 196.97	80 **Hg** 200.59	81 2.04 **Ti** 204.37	82 2.32 **Pb** 207.19	83 2.02 **Bi** 208.98	84 **Po** (210)	85 ***At*** (210)	86 **Rn** (222)
87 **Fr** (223)	88 **Ra** 226.025	89 **Ac** 227.0	104 **Rf** (261)	105 **Db** (262)	106 **Sg** (263)	107 **Bh**	108 **Hs**	109 **Mt**	110 **Uun**	111 **Uuu**	112 **Unb**						

58 **Ce** 140.12	59 **Pr** 140.91	60 **Nd** 144.24	61 **Pm** (147)	62 **Sm** 150.35	63 **Eu** 151.96	64 **Gd** 157.25	65 **Tb** 158.92	66 **Dy** 162.50	67 **Ho** 164.93	68 **Er** 167.26	69 **Tm** 168.93	70 **Yb** 173.04	71 **Lu** 174.97
90 **Th** 232.04	91 **Pa** (231)	92 **U** 238.03	93 **Np** (237)	94 **Pu** (242)	95 **Am** (243)	96 **Cm** (247)	97 **Bk** (247)	98 **Cf** (249)	99 **Es** (254)	100 **Fm** (253)	101 **Md** (253)	102 **No** (256)	103 **Lw** (260)

INDEX